T0245221

CAMBRIDGE LIBRARY COLLECTION

Books of enduring scholarly value

Earth Sciences

In the nineteenth century, geology emerged as a distinct academic discipline. It pointed the way towards the theory of evolution, as scientists including Gideon Mantell, Adam Sedgwick, Charles Lyell and Roderick Murchison began to use the evidence of minerals, rock formations and fossils to demonstrate that the earth was older by millions of years than the conventional, Bible-based wisdom had supposed. They argued convincingly that the climate, flora and fauna of the distant past could be deduced from geological evidence. Volcanic activity, the formation of mountains, and the action of glaciers and rivers, tides and ocean currents also became better understood. This series includes landmark publications by pioneers of the modern earth sciences, who advanced the scientific understanding of our planet and the processes by which it is constantly re-shaped.

Discussions on Climate and Cosmology

The cause of the ice ages was a puzzle to nineteenth-century climatologists. One of the most popular theories was that the affected continents must somehow have been hugely elevated and, like mountains, iced over. However, in this 1885 study of the problem, James Croll (1821–90) argues that such staggering movement would have been impossible. Instead, he puts forward a new theory: that the eccentricity of the earth's orbit changes at regular intervals over long periods, creating 'great secular summers and winters'. Adopting a meticulous approach to the facts, he disproves a host of well-established notions across several disciplines and makes some remarkable deductions, including the effect of ocean currents on climate, the temperature of space, and even the age of the sun. With a focus on logical argument and explanation rather than mathematics, his book remains fascinating and accessible to students in the history of science.

Cambridge University Press has long been a pioneer in the reissuing of out-of-print titles from its own backlist, producing digital reprints of books that are still sought after by scholars and students but could not be reprinted economically using traditional technology. The Cambridge Library Collection extends this activity to a wider range of books which are still of importance to researchers and professionals, either for the source material they contain, or as landmarks in the history of their academic discipline.

Drawing from the world-renowned collections in the Cambridge University Library and other partner libraries, and guided by the advice of experts in each subject area, Cambridge University Press is using state-of-the-art scanning machines in its own Printing House to capture the content of each book selected for inclusion. The files are processed to give a consistently clear, crisp image, and the books finished to the high quality standard for which the Press is recognised around the world. The latest print-on-demand technology ensures that the books will remain available indefinitely, and that orders for single or multiple copies can quickly be supplied.

The Cambridge Library Collection brings back to life books of enduring scholarly value (including out-of-copyright works originally issued by other publishers) across a wide range of disciplines in the humanities and social sciences and in science and technology.

Discussions on Climate and Cosmology

James Croll

CAMBRIDGE UNIVERSITY PRESS

Cambridge, New York, Melbourne, Madrid, Cape Town,
Singapore, São Paolo, Delhi, Mexico City

Published in the United States of America by Cambridge University Press, New York

www.cambridge.org
Information on this title: www.cambridge.org/9781108055307

This edition first published 1885
This digitally printed version 2013

ISBN 978-1-108-05530-7 Paperback

DISCUSSIONS

ON

CLIMATE AND COSMOLOGY

DISCUSSIONS

ON

CLIMATE AND COSMOLOGY

BY

JAMES CROLL LL.D. F.R.S.

AUTHOR OF "CLIMATE AND TIME" "PHILOSOPHY OF THEISM" &c

EDINBURGH

ADAM AND CHARLES BLACK

MDCCCLXXXV

PREFACE.

ONE object of the present volume—the appearance of which has been long delayed by circumstances over which I had no control—is to consider the objections which have been urged from time to time against the Physical Theory of Secular Changes of Climate advanced in my work *Climate and Time*, and in previous papers on the subject. Most probably I would not have replied to my critics had it not been that the examination of their objections afforded an excellent opportunity for discussing more fully some comparatively obscure and difficult points in Geological Climatology. Such discussion is the main and primary object I have in view in the present volume.

A chapter or two have been devoted to a fuller consideration of the physical conditions of Continental Ice than is to be found in *Climate and Time*. I have also treated at greater length the question of the cause of Mild Polar Climates, and likewise the growing mass of evidence which we have in support of Arctic Interglacial Periods.

In one of the chapters towards the end of the volume, the reader will find evidence from the testimony of geology that the age of the sun's heat must be far greater than it could possibly have been had the heat been derived, as is generally supposed, from the condensation of the sun's mass.

I may mention that although the greater part of the work has already appeared in separate articles in the *Philosophical Magazine* and some other journals, the volume, nevertheless, is not a collection of detached papers; for nearly the whole of the articles were written in connected order with the view of appearing in their present form.

There are many of the topics discussed which I could have wished to consider more at length, but advancing years and declining health constrain me to husband my remaining energies for work in a wholly different field of inquiry—work which has never lost for me its fascination, but which has been laid aside for upwards of a quarter of a century.

EDINBURGH, *October*, 1885.

CONTENTS.

CHAPTER I.

THE FAILURE OF ATTEMPTS TO ACCOUNT FOR SECULAR CHANGES OF CLIMATE.

PAGE

Most important Problem in Terrestrial Physics.—Early Attempts to explain Geological Climate.—Lyell's Theory.—Theory of Change in Obliquity of Ecliptic.—Theory of Change in Axis of Rotation.—Tendency in Geology to Cataclysmic Theories. —The True Theory not Cataclysmic.—An Important Difference, ... 1

CHAPTER II.

MISAPPREHENSIONS REGARDING THE PHYSICAL THEORY OF SECULAR CHANGES OF CLIMATE.—REPLY TO CRITICS.

Reason for considering Professor Newcomb's Objections.—Logical Analysis rather than Mathematics required in the meantime. —Temperature of Space.—Law of Dulong and Petit.—Heat Conveyed by Aërial Currents.—Why the Mean Temperature of the Ocean ought to be greater than that of the Land.—Heat Cut Off by the Atmosphere, ... 17

CHAPTER III.

MISAPPREHENSIONS REGARDING THE PHYSICAL THEORY OF SECULAR CHANGES OF CLIMATE.—REPLY TO CRITICS.—*Continued.*

Tables of Eccentricity.—Influence of Winter in Aphelion.—Influence of a Snow-covered Surface.—Heat Evolved by Freezing. —The Fundamental Misconception.—The Mutual Reaction of the Physical Agents.—Explanation begins with Winter. —Herr Woeikof on the Cause of Glaciation, ... 38

CHAPTER IV.

OBJECTION THAT THE AIR AT THE EQUATOR IS NOT HOTTER IN JANUARY THAN IN JULY.

PAGE

Influence of the present distribution of Land and Water.—The Summer of the Southern Hemisphere colder than that of the Northern.—Influence of the Trade Winds on the Temperature of the Equator, ... 59

CHAPTER V.

THE ICE OF GREENLAND AND THE ANTARCTIC CONTINENT NOT DUE TO ELEVATION OF THE LAND.

Greenland; attempts to Penetrate into the Interior.—No Mountain Ranges in the Interior.—The Föhn of Greenland.—Antarctic Regions.—Character of the Icebergs.—Sir Joseph Dalton Hooker and Professor Shaler on the Antarctic Ice.—On the Argument against the Existence of a South-Polar Ice-Cap.—Thickness of Ice not dependent on Amount of Snowfall, ... 64

CHAPTER VI.

EXAMINATION OF MR. ALFRED R. WALLACE'S MODIFICATION OF THE PHYSICAL THEORY OF SECULAR CHANGES OF CLIMATE.

Effect of Winter Solstice in Aphelion.—The Star Storage of Cold.—Highland and heavy Snowfall in Relation to the Glacial Epoch.—The only Continental Ice on the Globe probably on Lowlands.—Modification of Theory Examined.—General Statement of the Theory.—Points of Agreement, . 82

CHAPTER VII.

EXAMINATION OF MR. ALFRED R. WALLACE'S MODIFICATION OF THE PHYSICAL THEORY OF SECULAR CHANGES OF CLIMATE. —*Continued.*

PAGE

Physics in relation to Mr. Wallace's Modification of the Theory. —Another Impossible Condition Assumed.—A Geographical Change not Necessary in Order to Remove the Antarctic Ice.—Other Causes than Antarctic Ice affecting the Northward-flowing Currents.—Climatic Conditions of the Two Hemispheres the reverse Ten Thousand Years ago; argument from.—Mutual Reactions of the Physical Agents in Relation to the Melting of the Ice.—Another Reason why the Ice does not Melt, ... 100

CHAPTER VIII.

EXAMINATION OF MR. ALFRED R. WALLACE'S MODIFICATION OF THE PHYSICAL THEORY OF SECULAR CHANGES OF CLIMATE. —*Continued.*

Professor J. Geikie on Condition of Europe during Interglacial Periods.—Scotland during Interglacial Periods.—Difficulty in Detecting the Climatic Character of the Earlier Interglacial Periods.—Objection as to the Number of Interglacial Periods.—Objection as to the Number of Submergences.—Interglacial Periods less strongly marked in Temperate Regions than Glacial, ... 126

CHAPTER IX.

THE PHYSICAL CAUSE OF MILD POLAR CLIMATES.

The Probably True Explanation.—Sir William Thomson on Mild Arctic Climates.—Mr. Alfred R. Wallace on Mild Arctic Climates.—Influence of Eccentricity during the Tertiary Period, ... 143

CHAPTER X.

THE PHYSICAL CAUSE OF MILD POLAR CLIMATES.—*Continued.*

PAGE

Climate of the Tertiary period, in so far as affected by Eccentricity.—Evidence of Alternations of Climate.—Were there Glacial Epochs during the Tertiary period?—Evidence of Glaciation during the Tertiary period, ... 160

CHAPTER XI.

INTERGLACIAL PERIODS IN ARCTIC REGIONS.

Interglacial Periods in Arctic Regions more marked than Glacial. —Evidence from the Mammoth in Siberia. — Northern Siberia much Warmer during the Mammoth Epoch than now.—Evidence from Wood.—Evidence from Shells.—The Mammoth Interglacial.—Main Characteristics of Interglacial Climate.—Evidence from the Mammoth in Europe. —The Mammoth Glacial as well as Interglacial.—Arctic America during Interglacial Times.—Was Greenland Free from Ice during any of the Interglacial Periods? . . 176

CHAPTER XII.

THE DISTRIBUTION OF FLORA AND FAUNA IN ARCTIC REGIONS.

Flora and Fauna of Iceland and the Faröe Islands destroyed by last Ice-sheet.—How was the present Flora and Fauna of these Islands introduced?—Professor J. Geikie's Explanation.—Ice and Ocean Currents as Transporting Agents, . 197

CHAPTER XIII.

PHYSICAL CONDITIONS OF THE ANTARCTIC ICE-SHEET.

Sir Wyville Thomson on the Antarctic ice.—Testimony of iceberg.—Temperature of the Antarctic ice.—Heat derived from beneath.—Heat derived from the upper surface.—Heat derived from work by compression and friction.—Temperature of the ice determined by the temperature of the surface.—Temperature of the ice in some regions determined by pressure, ... 202

CHAPTER XIV.

PHYSICAL CONDITIONS OF THE ANTARCTIC ICE-SHEET.—*Continued.*

PAGE

Limit to Thickness of the Ice resulting from Melting produced by Pressure.—Supposed Diminution in Thickness of Ice-strata from Compression and Melting.—Centre of Dispersion.—Ice thickest at Centre of Dispersion.—Thickness at Pole independent of Amount of Snowfall.—Rate of Motion of the Antarctic Ice.—Probable Thickness at the Pole.—Ice of the Glacial Epoch, ... 224

CHAPTER XV.

REGELATION AS A CAUSE OF GLACIER MOTION.

Why the Problem of Glacier Motion is so difficult.—Heat in Relation to Glacier Motion.—Regelation as a Cause of Motion.—Theories of the Cause of Regelation.—How Regelation produces Motion.—Heat transformed into Glacier Motion, 248

CHAPTER XVI.

THE TEMPERATURE OF SPACE AND ITS BEARING ON TERRESTRIAL PHYSICS.

The Importance of Knowing the Temperature of Space.—The Researches of Pouillet and Herschel in reference to the Temperature of Space.—A Defect in Dulong's and Petit's Formula.—Professor Balfour Stewart on Radiation of Thin Plates.—Radiation of Gases, ... 258

CHAPTER XVII.

THE PROBABLE ORIGIN AND AGE OF THE SUN'S HEAT.

Age of the Sun's Heat according to the Gravitation Theories.—Testimony of Geology as to the Age of Life on the Globe.—Evidence from "Faults."—Rate of Denudation.—Age of the Stratified Rocks as determined by the Rate of Denudation, 264

CHAPTER XVIII.

THE PROBABLE ORIGIN AND AGE OF THE SUN'S HEAT.—*Continued.*

PAGE

Not obliged to assume that Gravitation is the only Source of the Sun's Heat.—How the Mass obtained its Temperature.—No Limit to the Amount of Heat which may have been produced.—Age of the Sun in Relation to Evolution.—Note on Sir William Thomson's Arguments for the Age of the Earth, ... 282

CHAPTER XIX.

THE PROBABLE ORIGIN OF NEBULÆ.

Motion in space as a source of heat.—Reason why nebulæ occupy so much space.—Reason why nebulæ are of such various shapes.—Reason why nebulæ emit such feeble light.—Heat and light of nebulæ cannot result from condensation.—The gaseous state the first condition of a nebula.—Star-clusters.—Objections considered. 297

INDEX, ... 316

Chart of Probable Path of the Ice in North-Western Europe *To face page* 133.

CHAPTER I.

THE FAILURE OF ATTEMPTS TO ACCOUNT FOR SECULAR CHANGES OF CLIMATE.

Most important Problem in Terrestrial Physics.—Early Attempts to explain Geological Climate.—Lyell's Theory.—Theory of Change in Obliquity of Ecliptic.—Theory of Change in Axis of Rotation.—Tendency in Geology to Cataclysmic Theories.—The True Theory not Cataclysmic.—An Important Difference.

THE most important problem in terrestrial physics, in so far as regards geological and palæontological science, and the one which will ultimately prove the most far-reaching in its consequences, is, What are the physical causes which led to the Glacial Epoch and to all those great secular changes of climate which are known to have taken place during geological ages? How are we to account for the cold and Arctic condition of things which prevailed in temperate regions during what is called the Glacial Epoch, or for the warm and temperate climate enjoyed by the Arctic regions, probably up to the Pole, during part of the Miocene and other periods? Theories of the cause of those changes, of the most diverse and opposite character, have been keenly advocated, and one important result of the discussions which have recently taken place is the narrowing of the field of inquiry and the bringing of the question within proper limits.

The warm character of the climate of former ages was at one time generally referred by geologists to the influence of the earth's internal heat. But it was soon proved by Mr. Hopkins, Sir William Thomson, and others that this theory was utterly untenable. Sir W. Thomson, for example, showed that the general character of the climate of our globe could not have been sensibly affected by internal heat at any time beyond 10,000 years after the commencement of the solidification of the surface.

About twenty years ago an ingenious modification of the theory of internal heat was propounded by Professor Frankland.* He assumed that the changes of climate experienced by our earth during past epochs is to be referred to a difference in the influence of internal heat on the sea and on land. He concluded that the cooling of the floor of the ocean would proceed less rapidly than it would have done had it been freely exposed to the air, and that hence it would continue at a comparatively high temperature long after the surface of the dry land had reached its present mean temperature. And, as heat is transmitted from the bottom to the surface of the ocean, not by conduction, but by convection, *i.e.*, by the warm stratum of water in contact with the bottom rising to the surface, the temperature of the ocean would consequently be higher than the mean temperature of the earth's surface. He concluded that this state of things satisfactorily accounts for the Glacial Epoch. The higher temperature of the ocean would give rise to augmented atmospheric precipitation, and this would produce such an accumulation of snow during winter as would defy the heat of summer to melt.

This theory never seemed to gain acceptance amongst geologists, for it is well known that the sea of the Glacial Epoch was intensely cold—not warm.

* "Philosophical Magazine," May, 1864.

Others, again, tried to explain the great changes of climate by supposing, with M. Poisson, that the earth during its past geological history may have passed through hotter and colder parts of space. This theory, however, was found to be so much opposed to known physical principles that it soon had to be abandoned.

It has also been suggested that the changes of climate during geological ages might have resulted from alternations in the composition of the atmosphere, particularly in variations in the amount of carbonic acid gas possessed by the air.

A theory was advanced many years ago by Professor H. Le Coq that the changes of climate may have been due to changes in the amount of the sun's heat. But this theory, like the above, appears to have gained but little acceptance.

According to others, elevation of the land in the regions glaciated is assigned as the cause of that glaciation, and if the ice had been merely local, such an explanation might have sufficed. But we know the whole Northern Hemisphere, down to tolerably low latitudes, has been subjected in post-tertiary times to the rigour of an Arctic climate; so that, according to this theory, we must assume an upheaval of the entire hemisphere—an assumption too monstrous to be admitted, and as useless as absurd.

At one time Lyell's theory of the relative distribution of land and water was generally regarded by geologists as sufficient. It is, however, now generally admitted to be wholly insufficient to explain the now-known facts, and the conviction is becoming almost universal that we must refer the climatic changes in question to some cosmical cause.

The theory of a change in the obliquity of the ecliptic has been appealed to. This theory for a time met with a favourable reception, but, as might have been expected, it was soon abandoned. The researches of Mr. Stockwell of America, and of Professor George Darwin and others in this country, have put it beyond doubt that no probable amount of geographical revolution could ever have altered the obliquity to any sensible extent beyond its present narrow limits. It has been demonstrated, for example, by Professor George Darwin, that supposing the whole equatorial regions up to lat. 45° N. and S. were sea, and the water to the depth of 2000 feet were placed on the Polar regions in the form of ice—and this is the most favourable redistribution of weight possible for producing a change of obliquity—it would not shift the Arctic circle by so much as an inch!

Variations in the obliquity of the ecliptic having been given up as hopeless, geologists and physicists are now inquiring whether the true cause may not be found in a change in the position of the earth's axis of rotation. Fortunately this question has been taken up by several able mathematicians, among whom are Sir Wm. Thomson,* Professor Haughton,† Professor George Darwin,‡ the Rev. J. F. Twisden,§ and others; and the result arrived at ought to convince every geologist how hopeless it is to expect aid in this direction.

Professor George Darwin has demonstrated that in order to displace the pole merely 1° 46′ from its present position, $\frac{1}{20}$ of the entire surface of the globe would require to be elevated to a height of 10,000 feet, with a corresponding subsidence in another quadrant.

* "British Association Report," 1876 (part 2), p. 11.

† "Proceedings of Royal Society," vol. xxvi., p. 51.

‡ "Transactions of Royal Society," vol. 167 (part 1).

§ "Quart. Journ. Geol. Soc.," February, 1878.

There probably never was an upheaval of such magnitude in the history of our earth. And to produce a deflection of 3° 17′ (a deflection which would hardly sensibly affect climate) no less than $\frac{1}{10}$ of the entire surface would require to be elevated to that height. A continent ten times the size of Europe elevated two miles would do little more than bring London to the latitude of Edinburgh, or Edinburgh to the latitude of London. He must be a sanguine geologist indeed who can expect to account for the glaciation of this country, or for the former absence of ice around the poles, by this means. We know perfectly well that since the Glacial Epoch there have been no changes in the physical geography of the earth sufficient to deflect the Pole half-a-dozen miles, far less half-a-dozen degrees. It does not help the matter much to assume a distortion of the whole solid mass of the globe. This, it is true, would give a few degrees additional deflection of the Pole; but that such a distortion actually took place is more opposed to geology and physics than even the elevation of a continent ten times the size of Europe to a height of two miles.

Mr. Twisden, in his valuable memoir referred to, has shown even more convincingly how impossible it is to account for the great changes of geological climate on the hypothesis of a change in the axis of rotation. This conclusion has been further borne out by another mathematician, the Rev. E. Hill, in an article in the "Geological Magazine," for June, 1878. And Professor Haughton, in a paper read before the Royal Society, April 4, 1878, published in "Nature," July 4, 1878, entitled, "A Geological Proof that the Changes of Climate in Past Times were not due to Changes in the Position of the Pole," has proved from geological evidence that the Pole has never shifted its position to any great extent. "If we examine," he says, "the localities of the fossil remains of the Arctic regions, and consider carefully their relations to the position of the present North Pole, we find that we can demonstrate that the Pole has not sensibly changed its place during geological periods, and that the hypothesis of a shifting pole (even if permitted by mechanical considerations) is inadmissible to account for changes in geological climates."

There is no geological evidence to show that, at least since Silurian times, the Atlantic and Pacific were ever in their broad features otherwise than they are now—two immense oceans separated by the Eastern and Western continents—and there is not the shadow of a reason to conclude that the poles have ever shifted much from their present position. On this point I cannot do better than quote the opinion expressed by Sir William Thomson:—

"As to changes of the earth's axis, I need not repeat the statement of dynamical principles which I gave with experimental illustrations to the Society three years ago; but may remind you of the chief result, which is that, for steady rotation, the axis round which the earth revolves must be a 'principal axis of inertia,'—that is to say, such an axis that the centrifugal forces called into play by the rotation balance one another. The vast transpositions of matter at the earth's surface, or else distortions of the whole solid mass, which must have taken place to alter the axis sufficiently to produce sensible changes of the climate in any region, must be considered and shown to be possible or probable before any hypothesis accounting for changes of climate by alterations of the axis can be admitted. This question has been exhaustively dealt with by Professor George Darwin, in a paper recently communicated to the Royal Society of London, and the requisitions of dynamical mathematics for an alteration of even as much as two or three degrees in the earth's axis in what may be practically called geological time shown to be on purely geological grounds exceedingly improbable. But even suppose such a change as would bring ten or twenty degrees of more indulgent sky to the American Arctic Archipelago, it would bring Nova Zembla and Siberia by so much nearer to the pole; and it seems that there is probably as much need of accounting for a warm climate on one side as on the other side of the pole.* There is in fact no evidence in geological climate throughout those parts of the world which geological investigation has reached, to give any indication of the poles having been anywhere but where they are at any period of geological time."†

In the memoir from which the preceding paragraph is quoted, Sir William maintains that an increase in the amount of heat conveyed by ocean currents to the Arctic regions, combined with the effect of Clouds, Wind, and Aqueous Vapour, is perfectly sufficient to account for the warm and temperate condition of climate which is known to have prevailed in those regions during the Miocene and other periods.

Now, this is the very point for which I have been contending for many years. The only essential difference between Sir William's views and mine is simply this: He accounts for an increase in the flow of warm water to the Arctic regions by a submergence of the circumpolar land, whereas I attribute it to certain agencies brought into operation by an increase in the eccentricity of the earth's orbit. Such geological evidence as we possess of warm episodes in the

* This has been proved to be the case by Professor Haughton, "Nature," July 4, 1878.

† "Trans. of Geol. Soc. of Glasgow," Feb. 22, 1877.

polar regions does not point to such high temperatures being specially due to submergence of the polar land. What has chiefly tended to retard the acceptance of the theory of secular changes of climate discussed in my work entitled 'Climate and Time,' is the fact that physicists have not fully realised to what an immense extent the climatic condition of our globe is dependent upon the distribution of heat by means of ocean currents. Were it not for the enormous amount of heat transferred from equatorial to temperate and polar regions by means of ocean currents, the globe would scarcely be habitable by the present orders of sentient beings. When this fact becomes fully recognised, all difficulties felt in accounting for geological climate will soon disappear. The climatic influence of ocean currents has not been sufficiently considered, owing doubtless to the fact that before I attempted to compute the absolute amount of heat conveyed by the Gulf Stream, so as to compare it with the amount directly received by the Atlantic from the sun, no one had ever imagined that that ocean in temperate and Arctic regions was dependent to such an extent on heat brought from the Equator. And this being so, it was impossible for any one fully to realise to what an extent climate must necessarily be affected by an increase or a decrease of that stream.

Sir William Thomson speaks of his theory being that of Lyell; but beyond the mere assumption of the submergence of the circumpolar land, the two theories have little in common. Indeed, no one who believes (as Sir William does) that the former warm climatic condition of the polar area was mainly due to a transference of heat from equatorial to Arctic regions by means of ocean currents can logically adopt Lyell's theory.* According to that eminent geologist, the temperature of the Arctic regions would be raised by the removal of the continents from polar and temperate regions to a position along the equator. But if the equatorial regions were occupied by land instead of water, the possibility of conveying heat to temperate and Arctic regions by means of ocean currents would be completely cut off. In fact, one of the most effectual ways of lowering the mean temperature of the globe would be to group the continents along the equator.

The surface of the ground at the equator becomes intensely heated by the solar rays, and this heat is radiated into space much more rapidly than it would be from a surface of water warmed under the same conditions. Again, the air in contact with the hot ground becomes more speedily heated than it would if it were in contact with water, and consequently the ascending current of air over the equatorial lands carries off a greater amount of heat than it could take away from a water-surface. Now, were the heat thus carried off to be transferred by means of the upper currents to high latitudes, and there employed in heating the earth, then it might to a considerable degree compensate for the absence of warm ocean currents. In such a case land at the equator might be nearly as well adapted as water for raising the temperature of the whole earth. We know very well, however, that the heat carried up by the ascending current at the equator performs little work of this kind, but on the contrary is almost wholly dissipated into the cold stellar space above. Thus, instead of warming the globe, this ascending current is in reality a most effectual means of getting rid of the heat received from the sun, and

* For a fuller statement of Sir Wm. Thomson's views, see chapter on the Cause of Mild Polar Climates.

thereby reducing the temperature. Since, then, the earth loses as well as gains the greater part of her heat in equatorial regions, it is there that the substance best adapted for preventing the dissipation of that heat must be distributed in order to raise the general temperature. Now, of all substances in nature, water seems to possess this quality in the highest degree; and, being a fluid, it is adapted by means of currents to carry the heat which it receives to every region of the globe.

It has been urged as an objection to any ocean-current theory that, while it provides the requisite amount of heat, it fails to remove the three or four months' darkness of an Arctic winter, which must have proved fatal to plants of the Miocene period. This objection seems, however, to have no foundation in fact. Sir Joseph Hooker stated to the Royal Society, at the close of the reading of Prof. George Darwin's paper, that palms and other plants brought from the tropics survived the winter in St. Petersburg without damage, though matted down in absolute darkness for more than six months; and he was of opinion that the want of sunlight during the Arctic winter would not be very prejudicial to the plants.

But a cause must be found as well for the cold of the Glacial Epoch as for the warm climate of the Arctic regions that obtained in Miocene times. According to Lyell the continents would require to be moved to high temperate and polar regions to bring about a glacial condition of things in Britain. But this is an assumption which the present state of geological science will hardly admit. It is perfectly certain that there have been no such vast revolutions in physical geography in post-Tertiary times.

Tendency in Geology to Cataclysmic Theories.—

There has always been in Geology a tendency to cataclysmic theories of causation; a proneness to attribute the grand changes experienced by the earth's crust to extraordinary causes. Geologists have only slowly become convinced that those changes were the effects of the ordinary agencies in daily operation around us. For example, hills were formerly supposed to be due to sudden eruptions and upheavals; valleys to subsidences, and deep river gorges to violent dislocations of the earth. All this is now changed, and geologists in general have become convinced that the main features of the earth's surface owe their existence to the silent, gentle, and continuous working of such influences as rain and rivers, heat and cold, frost and snow.

It is not difficult to understand why a belief in cataclysms should so long have prevailed, and geologists should have been so prone to assume the existence of extraordinary causes acting with great force. Geological phenomena come directly under the eye in all their magnitude, and consequently produce a powerful impression on the mind. The quiet and gentle operations of nature's ordinary agencies appear utterly inadequate to produce results so stupendous; and one naturally refers effects so striking to extraordinary causes. Beholding in a moment the effect, we forget that the cause has been in operation for countless ages.

We look, for example, at a gorge, perhaps a thousand feet in depth, with a small streamlet running along its bottom. Our first impression is that this enormous chasm has been formed by some earthquake or other convulsion of nature rending the rocks asunder; and it is only when we examine the chasm more minutely, and find that it has been actually excavated out of the solid rock, that we begin to see that the work has been done by running water. At first, however, we do not imagine that such a chasm can have been made by the streamlet in its present puny form. We conclude that in former ages a great river ran down the channel. We fail to give the element of time due influence in our speculations. We overlook the fact that the streamlet has been deepening its bed for perhaps millions of years. Why, London itself might have been built by one man had he been at work during all the time that the streamlet was cutting out its gorge. When such considerations cross the mind, every difficulty vanishes, and we feel satisfied that all the work has been performed by the streamlet.

The very same may be said in regard to the origin of hills, valleys, and other features of the earth's surface. Yet how difficult it is still to convince some geologists that our mountains have been formed, as a rule, not by eruptions and upheavals, but by the slow process of sub-aerial denudation.

Cataclysmic explanations of phenomena have to a large extent disappeared from the field of physical geology. But there is one department in which they still monopolize the field, viz., in that which treats of great climatic changes in former ages. Just as in physical geology great and imposing effects have been attributed to extraordinary causes, so in questions of geological climate vast vicissitudes have been referred to equally vast and unusual agencies.

We know that at a period comparatively recent almost the entire Northern hemisphere down to tolerably low latitudes was buried under snow and ice, the climate being perhaps as rigorous as that of Greenland at the present day. And we know further that at other periods, Greenland and the Arctic regions were not only to a large extent at least free from ice, but also enjoyed a climate as warm and genial as that of England. To attribute results so striking and stupendous to such commonplace agencies as ocean currents, winds, clouds, and aqueous vapour is at present considered to be little else than absurd. Extraordinary and imposing causes proportionate to the effects are therefore sought.

To account for the Glacial Epoch, for example, the land was at one time supposed to have stood much higher than at present. It was soon discovered, however, that the glaciation was much too general to be explained by such means. Many believed that it might be accounted for by assuming a displacement of the continents, but this hypothesis had likewise to be abandoned when it became known that no alteration in the position of our continents and ocean basins has taken place since the Glacial Epoch.

Others again imagined that some great change had probably taken place in the obliquity of the ecliptic so as to bring the Arctic circle down to beyond the latitude of England. And in order to bring this about what enormous upheavals were supposed to have occurred! It was soon, however, shown that no possible rearrangement of matter on our globe could materially affect the obliquity; and besides this, it was further pointed out that, even supposing the Arctic circle was by such means to be shifted down to our latitude, yet it would not bring an Arctic climate along with it, but the reverse. This hypothesis being in its turn abandoned, it was next assumed that the earth's axis of rotation must have been moved so as to carry our island up to the Arctic regions. But to shift the axis of rotation even so much as 3°, upheavals and subsidences of a magnitude hitherto unheard of in geological speculations had to be assumed. A change of 3°, however, being totally inadequate to account for the great changes of climate in question, earthquakes of sufficient power to break up the solid framework of the globe had to be called into operation, so as to cause a rearrangement of matter sufficient to produce a displacement of the pole to the extent required. The amount of distortion necessitated by this theory is so enormous that most of its advocates have recently abandoned it as hopeless.

But is there really after all any necessity for invoking the aid of agencies so extraordinary and gigantic? To carve a country, say like Scotland, out of hard Silurian rock into hill and dale and mountain ridges, thousands of feet in height, is certainly a more stupendous undertaking than simply to cover the same area with a sheet of ice. And if commonplace agencies like rain and rivers, frost and snow, can do the former, why may not such agencies as ocean currents, winds, clouds, and aqueous vapour be sufficient for the latter?

That geological climate should depend on the causes to which we refer cannot appear more improbable to the geologists of the present day than the inference that hills and valleys were formed by atmospheric agencies did to the geologists of the last generation. And there is little doubt that by the next generation the one conclusion will be as freely admitted as the other.

When a physicist so eminent as Sir Wm. Thomson expresses his decided opinion that the agencies in question are all that are necessary to remove the ice from the Arctic regions, and confer on them a mild and temperate climate, it is to be hoped that the day is not far distant when the climate controversy will be concluded. When the fact comes to be generally admitted by physicists that a great *increase* in the temperature and volume of the ocean currents flowing polewards is sufficient to prevent the accumulation of ice in the Arctic regions, it will then be allowed that we only require a great *decrease* in the volume and temperature of the currents in order to account for the former accumulations of ice on the temperate regions, or, in other words, to explain the occurrence of the Glacial Epoch. And when this position is reached, it will be seen that the whole depends upon a very simple cause, requiring neither the submergence nor the elevation of continents, nor any other great change in the physical geography of the globe.

When the eccentricity of the earth's orbit is at a high value, and the Northern winter solstice is in perihelion, agencies are brought into operation which make the S.E. trade winds stronger than the N.E., and compel them to blow over upon the Northern hemisphere as far probably as the Tropic of Cancer. The result is that all the great equatorial currents of the ocean are impelled into the Northern hemisphere, which thus, in consequence of the immense accumulation of warm water, has its temperature raised, and snow and ice to a great extent must then disappear from the Arctic regions. When the precession of the equinoxes brings round the winter solstice to aphelion, the condition of things on the two hemispheres is reversed, and the N.E. trades then blow over upon the Southern hemisphere, carrying the great equatorial currents along with them. The warm water being thus wholly withdrawn from the Northern hemisphere, its temperature sinks enormously, and snow and ice begin to accumulate in temperate regions. The amount of precipitation in the form of snow in temperate regions is at the same time enormously increased by the excess of the evaporation in low latitudes resulting from the nearness of the sun in perihelion during summer.

The final result to which we are therefore led is that those warm and cold periods which have alternately prevailed during past ages are simply the great secular summers and winters of our globe, depending as truly as the annual ones do upon planetary motions, and like them also fulfilling some important ends in the economy of Nature.

An important difference. — The physical theory differs from all the preceding in this important respect, viz., that it contains no hypothetical elements. All the causes are *real*; none hypothetical. The conclusions are all deduced either from known facts, or from admitted physical principles, and in no case are they based on hypotheses. Hypotheses will be found in my cosmological discussions, but none when I deal with climatological questions.

CHAPTER II.

MISAPPREHENSIONS REGARDING THE PHYSICAL THEORY OF SECULAR CHANGES OF CLIMATE—REPLY TO CRITICS.

Reason for considering Professor Newcomb's Objections.—Logical Analysis rather than Mathematics required in the meantime.—Temperature of Space.—Law of Dulong and Petit.—Heat Conveyed by Aerial Currents.—Why the Mean Temperature of the Ocean ought to be greater than that of the Land.—Heat Cut Off by the Atmosphere.

TWENTY-ONE years ago the theory was advanced that the Glacial Epoch was the result of a combination of physical agents brought into operation by an increase in the eccentricity of the earth's orbit. Few or no objections have been urged against what may be called the astronomical part of the theory; but the portions relating to these physical agencies, which are by far the most important part, have from time to time met with considerable opposition. Considering the newness of the subject, and the complex nature of many of these combinations of physical agencies, it would not be surprising if some of the original deductions in regard to them proved erroneous; but after long and careful reconsideration of the whole matter, I have not found reason to abandon any of them or alter them to any material extent.

The only class of objections urged against the theory which I have as yet considered at length are those relating to the cause of ocean-currents, and their influence on the distribution of heat over the globe;

and I think it will be admitted that the views which I have advocated on these points are now being generally, if not almost universally accepted.

But it is in reference to the influence of aqueous vapour, fogs, and clouds on the production and preservation of snow that the greatest diversity of opinion has prevailed. The object of the present chapter is to examine at some length the principal objections which have been advanced in regard to this part of the inquiry. I shall also take the present opportunity of discussing more fully some points on which I have been sometimes misunderstood, and which appear to have been treated rather too briefly on former occasions.

In the "American Journal of Science" for April, 1876, Professor Newcomb has done me the honour to review at some length my work, 'Climate and Time;' and as his article is mainly devoted to a criticism of my reasoning in regard to those very points to which I refer, I shall begin with an examination of his objections. One reason for entering at some length into an examination of Professor Newcomb's objections is the fact that they embrace to a large extent those which have been urged by reviewers in Great Britain. Some of his objections, however, as will be seen, are based upon a misapprehension of my reasoning. More recently, in an article in that Journal for January, 1884, and in the "Phil. Mag." for February, 1884, he has advanced other objections.

Professor Newcomb states that he has a want of confidence in anything short of a purely mathematical investigation of the subject. Of course, I fully concur with him as to the desirability of a "purely mathematical investigation of the subject." Such an investigation, however, is, I think, impossible at present.

In a question so complex and difficult as that of the cause of the Glacial Epoch, depending as it does on the consideration of so many different elements, some of which are but little understood, logical analysis rather than mathematics will require to be our instrument in the meantime. The question must first assume a clear, definite, and logical form before mathematics can possibly be applied to it.

It is objected that my language is wanting in quantitative precision—that I use such terms as "great," "very great," "small," "comparatively small," and so forth without any statement of the units of comparison relatively to which these expressions are employed. No one reasoning on the combined influence of a multitude of physical causes could well avoid the almost continual use of such terms. Besides, my critic forgets that in almost every case in which I use these terms numerical exactness is not attainable; and even if it were, it would, as a rule, be of little service, seeing that the conclusion generally depends on the simple fact that one quantity is less or greater than another; not on *how much* less or *how much* greater the one may be than the other. Although my arguments are logical, few writers, I venture to say, have done more than myself to introduce definite quantitative exactness into the questions I have discussed.

Temperature of Space.—One of the most important factors in the theory of geological climate resulting from changes in the eccentricity of the earth's orbit is obviously the temperature of stellar space. Unless we have, at least, some rough idea of the proportion which the heat derived from the stars bears to that derived from the sun, we cannot form any estimate of how much the temperature of our earth would be lowered or raised by a given decrease or increase of the sun's distance.

The question of the temperature of space has been investigated in different ways by Pouillet and Herschel; and the result arrived at was that space has a temperature of –239° F., or an absolute temperature of 222°. The mean absolute temperature of our earth is about 521°. Consequently according to these results, the heat received from the stars is to that received from the sun as 222 to 299. All my determinations of the change of temperature due to changes in the sun's distance were computed on these data, although I believe, for reasons stated, that space must have a much lower temperature. Recent observations of Professor Langley made during the Mount-Whitney expedition confirm the correctness of my belief.

Professor Newcomb, however, wholly ignores all that has been done on that subject, for he commences his review by the statement that "practically there is but one source from which the surface of the earth receives heat—the sun, since the quantity received from all other sources is quite insignificant in comparison."

He states that he regards the conclusion that the temperature of space is –239° as having no sound basis. This may be perfectly true; but it is hardly a warrant for affirming that practically there is but one source (the sun) from which the surface of the earth receives heat, without even referring to the researches of these eminent physicists who have arrived at a totally different conclusion. Any one who has read 'Climate and Time' will know that I adopted – 239° as the temperature of space, not because I believed that estimate to be correct, but because at the time I wrote there was no other to adopt. In fact in adopting so high a temperature for space I was doing my theory a positive injury. This is obvious; for the lower the temperature of space the greater must be the decrease of temperature resulting from an increase in the sun's distance due to eccentricity. My opinion all along has been that the temperature of space is little above absolute zero.

As an argument against the conclusion that space can have the high temperature assigned to it by Pouillet and Herschel, he says:—"Photometry shows that the combined light from all the stars visible in the most powerful telescope is not a millionth of that received from the sun, and there is no reason for believing that the ratio of light to heat is incomparably different in the two cases." This very argument from the extreme smallness in the amount of light derived from the stars in comparison to that from the sun, intended by him to convince me of the absurdity of supposing that space possesses a temperature as high as −239°, is just the argument advanced by myself in the "Reader" for December 9, 1865, and afterwards reproduced in 'Climate and Time,' at page 39, from which I quote the following:—

'We know that absolute zero is at least 493° below the melting-point of ice. This is 222° below that of space. Consequently, if the heat derived from the stars is able to maintain a temperature of −239°, or 222° of absolute temperature, then nearly as much heat is derived from the stars as from the sun. But if so, why do the stars give so much heat and so very little light? If the radiation from the stars could maintain a thermometer 222° above absolute zero, then space must be far more transparent to heat-rays than to light-rays, or else the stars give out a great amount of heat, but very little light, neither of which suppositions is probably true. The probability is, I venture to presume, *that the temperature of space is not very much above absolute zero.*'

Law of Dulong and Petit.—Prof. Newcomb gives his readers to understand that I assume Newton's laws of cooling to be correct; and that I apparently nowhere adduce the more correct law of Dulong and Petit—viz., that if we take a series of temperatures in arithmetical progression, the corresponding rates of radiation of heat will not be in arithmetical progression, but in a series of which the differences continually increase. If he will refer to the "Reader," Dec. 9, 1865, "Phil. Mag.," Feb. 1870, "Nature," April 1, 1880, and 'Climate and Time' (the book he reviewed), p. 37, he will there see the question discussed at considerable length. He will also find reference made to a remarkable circumstance connected with radiation which perhaps may be new to him. It is this: the law of Dulong and Petit (that as the temperature of a body rises the radiation of the body increases in a much higher ratio) holds true only of the body considered as a mass. The probability is, as has been shown by Prof. Balfour Stewart, that the individual particles composing the body obey Newton's law in their radiation; in other words, the radiation of a material particle is directly proportionate to its absolute temperature.

Further, in estimating the extent to which temperature is affected by a change in the sun's distance, Newton's law makes the extent too great; while the formula of Dulong and Petit, which is an empirical one, makes it, on the other hand, too small. This formula has been found to agree pretty closely with observation within ordinary limits, but it completely breaks down when applied to determine high temperatures. For example, it is found to give a temperature for the sun of only 2130° F., or not much above that of an ordinary furnace. It is probable also that it will equally break down when applied to very low temperatures, such as that of space.

I am very much pleased to find that Professor Newcomb draws a conclusion from Dulong and Petit's law favourable to my theory of the cause of the Glacial Epoch, which certainly did escape my notice. And it is a curious circumstance that Mr. Hill* has likewise deduced a conclusion even more favourable to glaciation than that of Professor Newcomb.

I am pleased to find that he agrees, in the main, with what has been advanced in 'Climate and Time' in reference to the heating power of ocean currents, and also as to their existence being due to the impulse of the winds. But he differs widely from me in regard to the heat conveyed by aerial currents.

On the Heat conveyed by Aerial Currents.—I stated that the quantity of heat conveyed by the air from equatorial to high temperate and polar regions is trifling in comparison with that conveyed by ocean currents; for the heated air rising off the hot ground of the equator, after ascending a few miles becomes exposed to the intense cold of the upper regions, and having to travel polewards for thousands of miles in those regions, it loses nearly all the heat which it brought from the equator before it can possibly reach high latitudes. To this Professor Newcomb objects as follows: "He (Mr Croll) speaks of the hot air rising from the earth and becoming exposed to the intense cold of the upper regions of the atmosphere But what can this cold be *but the coldness of the very air itself which has been rising up?* If the warm air rises up into the cold air, and becomes *cooled by contact* with the latter, the latter must become warm by the very heat which the former loses; and if there is a continuous rising current the whole region must take the natural temperature

* "Evaporation and Eccentricity as Co-factors in Glacial Periods," "Geological Magazine" for November, 1881.

of the rising air. This temperature is, indeed, much below that which maintains at the surface, *for the simple reason that air becomes cold by expansion* according to a definite and well-known law. Having thus got his rising current constantly cooled off *by contact with the cold air* of the upper regions, it has to pass on its journey towards the poles," etc. (p. 267).*

Here the cooling of the ascending air is attributed to two causes—(1) the heat lost by expansion as the air rises; (2) the heat lost by contact with the colder air through which the ascending air passes and with which it mixes in the upper regions. But the two may be resolved into one, viz., the heat lost by expansion; for the cold air, to which the ascending air communicates its heat by contact, is assumed to have originally derived its cold, in like manner, from expansion. This is evident, for, although he recognizes the effect of radiation into space, he assumes that this loss is compensated by counter-radiation. The upper regions are, he says, exposed to the radiation of the sun on the one side, and of the earth's lower atmosphere on the other, and there is no proof that these do not equal the surface-temperature. And again, when the air descends in high latitudes to the earth's surface, an amount of heat will be evolved by compression equal to that which is lost when it rose from the equator.

Professor Newcomb has misapprehended not only my meaning, but also the chief reason why the air in the upper region is so intensely cold. Any one who has read what I have stated in pp. 35-40, 'Climate and Time,' regarding the temperature of space will readily understand what I mean by the temperature of the upper regions. By the temperature of stellar space, it is not meant that space itself is a something

* The italics are mine.

possessed of a given temperature, say $-230°$ F. It simply means the temperature to which a body would fall were it exposed to no other source of heat than that of radiation from the stars. By the temperature of the upper regions I mean the temperature to which air in those regions sinks in consequence of loss from radiation into space. It is mainly to this cause, and not to the loss from expansion, as Professor Newcomb assumes, that the intense cold of the upper air is due. The air in that region has got beyond the screen which protected it when at the earth's surface, and it then throws off its heat into space during twelve hours of night, getting no return from without except from the radiation of the stars. And even at noonday, as I have endeavoured to show in Appendix to 'Climate and Time,' p. 551, the rays of a burning sun overhead would not be sufficient to raise the temperature of the air up to the freezing-point. But the recent observations of Professor Langley prove that the loss of heat from radiation is in reality far greater than I had anticipated. He says:—"The original observations, which will be given at length, lead to the conclusion that in the absence of an atmosphere the earth's temperature of insolation would at any rate fall below $-50°$ F.; by which it is meant that, for instance, mercury would remain a solid under the vertical rays of a tropical sun were radiation into space wholly unchecked, or even if, the atmosphere existing, it let radiations of all wave-lengths pass out as easily as they come in" ("Nature," August 3rd, 1882).

The temperature of the upper atmosphere, even after making allowance for heat received from below, must in this case be at least 80° below the freezing-point. The quantity of heat lost by expansion must therefore be trifling compared with that lost by radiation; and although the heat lost by expansion is fully restored by compression, yet the air would reach the earth deprived almost entirely of the heat with which it left the equator. All that it could possibly give back would simply be the heat of compression; and this would hardly be sufficient to raise the air at −50° F. to the freezing-point. How then can the polar regions be greatly the better of air from the equatorial regions? Professor Newcomb says:—"If the upper current be as great as is commonly supposed, it must be as powerful as ocean-currents in tending to equalise the temperature of the globe." How can this be?

Why the Mean Temperature of the Ocean should be greater than that of the Land.—"Another proposition," he says, "which the author attempts to prove, reasoning which seems equally inconclusive, is that the mean temperature of the ocean is greater than that of the land over the entire globe." I certainly never attempted to prove that the mean temperature of the ocean is *greater* than that of the land over the entire globe. The very chapter to which he here refers, and which he is about to criticise, was written to explain why the mean temperature of the southern or water hemisphere is less than that of the northern or land hemisphere. What I attempted to prove was, not that the mean temperature of the ocean *is* greater than that of the land, but that, were it not for certain causes, the mean temperature of the ocean *ought* to be greater than that of the land in equatorial regions as well as in temperate and arctic regions. In other words, the object of the chapter was to prove that the mean temperature of the southern or water hemisphere was less than that of the northern or land hemisphere, not, as was generally supposed, because the former is mainly water and the latter land, but because of the enormous amount of heat transferred from the former to the latter hemisphere by means of ocean-currents; and that, were it not for this transference, the temperature of the water would exceed that of the land hemisphere.* And it is in order to prove this that the "four *à priori* reasons" which Professor Newcomb criticises were adduced. The first of these is as follows:—

First.—'The ground stores up heat only by the slow process of conduction, whereas water, by the mobility of its particles and its transparency for heat-rays, especially those from the sun, becomes heated to a considerable depth rapidly. The quantity of heat stored up in the ground is thus comparatively small, while the quantity stored up in the ocean is great.' †

These sentences are considered unworthy of criticism. Are they really so unworthy? Let us examine them a little more closely. It is in consequence of the sun's rays being able to penetrate to a great depth that the amount of heat stored up by the ocean is so great; and it is to this store that its warmth during winter is mainly due. The water is diathermanous for the rays of the sun, but it is not so, for reasons well known, for the rays of water itself. The upper layers of the ocean will allow a larger portion of the radiation from the sun to pass freely downward, but they will not allow radiation from the layers underneath to pass freely upwards. These upper layers, like the glass of a greenhouse, act as a trap to the

* Since 'Climate and Time' was published it has been proved from observation (see next chapter) that, notwithstanding this transference of heat, the water hemisphere is the warmer of the two.

† 'Climate and Time,' p. 90.

sun's rays, and thus allow the water of the ocean to stand at a higher temperature than it would otherwise do. Again, the slowness with which the ocean thus parts with its heat enables it to maintain that comparatively high temperature during the long winter months. And again, it is to the mobility of the particles of water, the depth to which the heat penetrates, and the rapidity with which it is absorbed, that those great currents of warm water become possible. Were the waters of the ocean like the land, not mobile, and were only a few inches at the surface reached by heat from the sun, there could be no Gulf-stream, or any great transference of heat from the southern to the northern hemisphere, or from equatorial to temperate and polar regions, by means of oceanic circulation.

Second.—'The air is probably heated more rapidly by contact with the ground than with the ocean; but on the other hand, it is heated far more rapidly by radiation from the ocean than from the land. The aqueous vapour of the air is to a great extent diathermanous to radiation from the ground, while it absorbs the rays from water, and thus becomes heated.'

To this Professor Newcomb objects, as follows:—"If, then, the air is really heated by contact with the ground more rapidly than by contact with the ocean, it can only be because the ground is hotter than the ocean, which is directly contrary to the theory Mr. Croll is maintaining." What I maintained was that, were it not for certain causes, the *mean annual* temperature of the ocean would be higher than that of the land. During the day and also during the summer the surface of the ground is hotter than that of the ocean; and the air, of course, will be heated more rapidly by *contact* with the former than with the latter. But this does not prove that the air is not more rapidly heated by *radiation* from the ocean than from the land. Professor Newcomb says:—"The statement that the aqueous vapour of the air is diathermanous to radiation from land, but not to that from water, is quite new to us, and very surprising." I am surprised that he is not acquainted with the fact, and also with its physical explanation. This will help to account for his inability to perceive how radiation from the ocean may heat the air more rapidly than radiation from the land, even though the surface of the latter may be at a higher temperature than that of the former.

He says:—"The rapidity with which the heating process goes on depends on the difference of temperature, no matter whether the heat passes by conduction or by radiation." This statement will hardly harmonise with recent researches into radiant heat. It is found that the rapidity with which a body is heated by radiation depends upon the absorbing power of the body; and the absorbing power again depends upon the quality of the heat-rays. Professor Tyndall, for example, found that in the case of vapours, as a rule, absorption *diminishes* as the temperature rises. With a platinum spiral heated till it was barely visible, the absorption of the vapour of bisulphide of carbon was 6·5, but when the spiral was raised to a white heat the absorption, was reduced to 2·9. A similar result took place in the case of chloroform, formic ether, acetic ether, and other vapours. The physical cause of this is well known.

If the aqueous vapour of the air, he says, be more diathermanous to radiation from land than from water, as I have stated, then I assigned directly contrary effects to the same cause. For, "reasoning as in (1), he, Mr. Croll, would have said that the air over the land, owing to its transparency for the heat-rays from the land, becomes heated to a greater height rapidly, while the air over the ocean, not being transparent, can acquire heat from the ocean only by the slow process of convection." I would have said no such thing. Radiation from the surface of the land will, no doubt, penetrate more freely through the aqueous vapour than radiation from the ocean; but the aqueous vapour will not absorb the radiation of the land so rapidly as that of the ocean, for the ocean gives off that quality of rays which aqueous vapour absorbs most rapidly.

This is not in opposition to what I have stated in reason (1); for, if the ground were transparent to the sun's rays like water, evidently the total quantity of heat absorbed by it would be greater than that by the ocean. But radiation from the sun heats only the surface of the ground; all below the surface depends for its supply on the slow process of conduction, whereas the ocean is heated by direct radiation to great depths. Consequently the total quantity of heat absorbed by the ocean, say per square mile, in a given time, is greater than that absorbed by the land.

Third.—'The air radiates back a considerable portion of its heat, and the ocean absorbs this radiation from the air more readily than the ground does. The ocean will not reflect the heat from the aqueous vapour of the air, but absorbs it, while the ground does the opposite. Radiation from the air, therefore, tends more readily to heat the ocean than it does the land.'

"Here we have," he says, "the air giving back to the ocean the same heat which it absorbs from it, and thus heating it." If Professor Newcomb means by this same heat the same amount of heat, then I believe in no such thing. But if his meaning be that here we have the air giving back to the ocean a quantity of the heat which it absorbed from it, then he is certainly correct in supposing that this is affirmed by me. But this is a conclusion which no physicist could for a moment doubt. To deny this would be to contradict Prevost's well-known theory of exchanges. Did the air throw back to the ocean none of the heat which it derives from it, the entire waters of the ocean would soon become solid ice. In fact, as we have seen, mercury would not remain fluid, and every living thing on the face of the globe would perish.

In his "Rejoinder,"* Professor Newcomb still maintains that this involves the *reductio ad absurdum* of two bodies heating each other by their mutual radiation. This is not the state of the case at all, for both bodies receive their heat from the sun; their mutual radiation simply retains them at a higher temperature than they could otherwise have. Here Professor Newcomb appears to get into confusion owing to the meaning which he attributes to the word "heating." The views which I have advocated in reference to this mutual radiation are as follows:—According to the dynamical theory of heat, all bodies above absolute zero radiate heat. If we have two bodies, A at 200° and B at 400°, then, according to Prevost's theory of exchanges, A as truly radiates heat to B as B does to A. The radiation of A, of course, can never raise the temperature of B above 400°; but nevertheless the *tendency* of the radiation of A, in so far as it goes, is to raise the temperature of B. This is demonstrated by the fact that the temperature of B, in consequence of the radiation of A, is prevented from sinking so low

* "Amer. Jour. of Science," Jan., 1884; "Phil. Mag.," Feb., 1884

as it would otherwise do. All this is so well known to every student of thermodynamics, that I can hardly think Professor Newcomb, on reflection, will dispute its accuracy. And if he admits this, then he must also admit the soundness of my third reason, for this is the principle on which it is based. The aqueous vapour of the air absorbs a considerable amount of the heat which is being constantly radiated by the ocean; a portion of this heat thus absorbed is thrown back upon the ocean, the tendency of which is to keep the surface of the ocean at a much higher temperature than it would otherwise have. Professor Langley has concluded, from observations made at Mount Whitney, that were it not for the heat thrown back by the atmosphere, or "trapped," as it is popularly called, mercury would remain solid under a vertical sun.

He states that reason fourth seems to be little more than a repetition of reason second in a different form. It is, however, much more than that. It is a demonstration that, were it not for the causes to which I have alluded, the mean temperature of the water hemisphere ought to be higher than that of the land hemisphere; and for this reason I shall here give the section in full.

Fourth.—'The aqueous vapour of the air acts as a screen to prevent the loss by radiation from water, while it allows radiation from the ground to pass more freely into space; the atmosphere over the ocean consequently throws back a greater amount of heat than is thrown back by the atmosphere over the land. The sea in this case has a much greater difficulty than the land has in getting quit of the heat received from the sun; in other words, the land tends to lose its heat more rapidly than the sea. The consequence of all these circumstances is that the ocean must stand at a higher mean temperature than the land. A state of equilibrium is never gained until the rate at which a body receiving heat is equal to the rate at which it is losing it; but as equal surfaces of sea and land receive from the sun the same amount of heat, it therefore follows that in order that the sea may get quit of its heat as rapidly as the land, it *must stand at a higher temperature* than the land. The temperature of the sea must continue to rise till the amount of heat thrown off into space equals that received from the sun; when this point is reached, equilibrium is established and the temperature remains stationary. But, owing to the greater difficulty that the sea has in getting rid of its heat, the mean temperature of equilibrium of the ocean must be higher than that of the land; consequently, the mean temperature of the ocean, and also of the air immediately over it, in tropical regions, should be higher than the mean temperature of the land and the air over it.'

I had thought that the foregoing expressed, with sufficient clearness, the reasons why the ocean ought to be warmer than the land; but I find that Professor Newcomb, in his "Rejoinder," still maintains that my views on this point are opposed to the fundamental laws of thermodynamics. But surely he must have misapprehended my reasoning.

The temperature of a body can remain stationary only when the rate at which it is losing equals that at which it is receiving heat. If heat be lost more rapidly than it is received, the temperature will fall. The fall of temperature will diminish the rate of loss till the rate of loss equals the rate of gain. After this the temperature becomes stationary. If we have two bodies, A and B, the same in every respect, each receiving (say from the sun) the same amount of heat in a given time, and if the only difference between them be that A has a greater difficulty than B in getting quit of the heat which it is receiving, then, for the reason just assigned, A will necessarily stand at a higher temperature than B. Let us now suppose the southern, or water hemisphere, to be A, and the northern, or land hemisphere, to be B. I have endeavoured to show ('Climate and Time,' and elsewhere) that A, the water hemisphere, ought to have a higher mean temperature than B, the land hemisphere, because the former has a greater difficulty in getting quit of the heat which it is receiving from the sun than the latter. The question then arises, How is it that the water hemisphere has a greater difficulty than the land hemisphere in getting rid of its heat? It is mainly due to that cause which Professor Newcomb says is quite new to him, viz. the fact that *the aqueous vapour of the air is far less diathermanous to radiation from water than from land.* It is a curious fact that Prof. Newcomb, in his "Rejoinder," entirely overlooks this cause assigned by me, although I have stated it fully in my fourth reason. The *period* of the heat-vibrations of the aqueous vapour of the air is the same as that of the ocean, and consequently the aqueous vapour will absorb radiation from the ocean more readily than from the land. A considerable portion of the heat absorbed by the aqueous vapour of the air is thrown back upon the ocean, and in this way the aqueous vapour acts as a screen, or like the glass of a greenhouse, in preventing the ocean from getting quit of its heat so rapidly as the land. The result is that the temperature of equilibrium of the ocean must be higher than that of the land. In other words, before the ocean can manage to throw off its heat into space as rapidly as it is receiving it, its temperature must be higher than that of the land.

The foregoing conclusion follows so obviously from the known properties of aqueous vapour and the principles of thermodynamics that I can hardly believe Prof. Newcomb will call it in question. But he will ask, How can the transparency of the ocean for heat-rays, the mobility of its particles, and the greater store of heat which it possesses, be a reason why its mean temperature should be higher than that of the land? I thought I had made all this clear. The reason becomes apparent when we consider why it is that the surface of the ocean during night, and also during winter, is warmer than the surface of the land. The ocean in temperate regions seldom sinks to the freezing-point, while the land is frequently frozen for months. The cause is obvious enough: at night, when the surface of the ocean becomes cool, the cold particles sink and their places are supplied by warm particles from below, and so long as the heat stored up remains, the surface can never become cold. Were it not for the transparency of water for heat-rays, it would be impossible that the ocean could obtain a supply of heat sufficient to maintain its surface-temperature during the entire winter; and, on the other hand, were the particles not mobile, this store could be of little service.

It is true that the land is hotter during the day, and also during the summer, than the ocean, but it is found that the more equable temperature of the ocean gives a higher mean. This is further shown from another consideration. The land is more indebted for heat to the ocean than the ocean is for heat to the land. For example, a very considerable portion of the warmth enjoyed by north-western Europe is derived from the Atlantic. In like manner, western America is indebted to the Pacific for a large amount of its heat. In addition, an immense quantity of the heat received from the sun by the ocean is consumed in producing evaporation, and a large portion of this heat latent in the vapour is bestowed on the land during condensation. Yet notwithstanding this transference of heat from the ocean to the land, the mean temperature of the former is greater than that of the latter. Were it not for its store of summer heat, the ocean could not afford to part with so much of its heat to the land during winter, and still maintain a higher mean temperature.

Since the publication of 'Climate and Time,' the accuracy of this conclusion has been confirmed in a remarkable manner from more recent researches on the actual mean temperature of the two hemispheres, the details of which have been given by Mr. Ferrell in his "Meteorological Researches" (Washington, 1877). It is found that the mean temperature of the northern or land hemisphere is higher than that of the southern or water hemisphere up only to about latitude 35°, and that beyond this latitude the mean temperature of the water hemisphere is the greater of the two. At latitude 40° the mean temperature of the southern hemisphere is 1°·4 higher than that of the same parallel on the northern hemisphere. At latitude 50° the difference amounts to 4°·4; while at latitude 60° the mean temperature of the southern hemisphere is actually 6° higher than that of the northern on the same parallel. The mean temperatures of the two hemispheres are as follows:—

Lat.	0°	10°	20°	30°	40°	50°	60°	70°	80°
Northern,...	80°·1	81°·0	77°·6	67°·6	56°·5	43°·4	29°·3	14°·4	4°·5
Southern,...	80·1	78·7	74·7	66·7	57·9	47·8	35·3		

From the above table we see that it is only in that area lying between the equator and latitude 35° that the southern hemisphere has a lower mean temperature than the northern. But it is from this area that the enormous amount of heat transferred to the northern hemisphere is mainly derived. Were the transference of heat to cease, the temperature of this area would be very considerably raised, and that of the corresponding area on the northern hemisphere lowered. The result would doubtless be that the southern hemisphere down to the equator would then be warmer than the northern. But, even as things are, as Mr. Ferrel remarks, "the mean temperature of the southern hemisphere is the greater of the two," the mean temperature of the southern being 60°·89 F. and that of the northern 59°·54 F.

Heat cut off by the Atmosphere.—Professor Newcomb says further, "Another idea of the author which calls for explanation is that solar heat absorbed by the atmosphere is entirely lost, so far as warming any region of the globe is concerned." This is no idea of mine. My idea is not that the heat cut off is entirely lost, but merely that the *greater part* is lost. A large portion of the heat is reflected, and of that absorbed one half, perhaps, is radiated back into space and lost, in so far as the earth is concerned.

CHAPTER III.

MISAPPREHENSIONS REGARDING THE PHYSICAL THEORY OF SECULAR CHANGES OF CLIMATE.—REPLY TO CRITICS—*Continued.*

Tables of Eccentricity.—Influence of Winter in Aphelion.—Influence of a Snow-covered Surface.—Heat Evolved by Freezing.—The Fundamental Misconception.—The Mutual Reaction of the Physical Agents.—Explanation begins with Winter.—Herr Woeikof on the Cause of Glaciation.

Tables of Eccentricity.—Referring to my tables of eccentricity of the earth's orbit, he says:—"That there are from time to time such periods of great eccentricity is a well-established result of the mutual gravitation of the planets; but whether the particular epochs of great and small eccentricity computed by Mr. Croll are reliable is a different question." I may here mention that Professor McFarland, of the Ohio State University, Columbus, a few years ago, undertook the task of re-computing every one of the 150 periods given in my tables, and he states that, except in one instance, he did not find an error to the amount of ·001.*

"The data for this computation," continues Professor Newcomb, "are the formulæ of Le Verrier, worked out about 1845,† without any correction either for the later corrections to the masses of the planets or for

* "American Journal of Science," vol. xi. p. 456 (1876).

† Le Verrier's formulæ were worked out several years before 1845.

the terms of the third order, subsequently discussed by Le Verrier himself. The probable magnitude of these corrections is such that reliance cannot be placed upon the values of eccentricity computed without reference to them for epochs distant by merely a million of years."

In regard to this objection I may mention that the whole subject of the secular variations of the elements of the planetary orbits has been re-investigated by Mr. Stockwell, taking into account the disturbing influence of the planet Neptune, the existence of which was not known at the time Le Verrier's investigations were made. Professor McFarland, with the aid of Mr. Stockwell's formulæ, has computed all the periods in the tables referred to above; and on comparing the results found by both formulæ, he states that "the two curves exhibit a general conformity throughout their whole extent." And his computations, I may state, extend from 3,260,000 years before 1850, and to 1,260,000 years after that date; or, in other words, over a period of no fewer than 4,520,000 years,* thus showing that Professor Newcomb's objection falls to the ground.

Influence of Winter in Aphelion.—I have maintained that at a time when the eccentricity is high and the winter occurs in aphelion, the great increase in the sun's distance and in the length of the winter would have the effect of causing a large increase in the quantity of snow falling during that season. This very obvious result follows as a necessary consequence from the fact that the moisture which now falls in the form of rain would

* In this laborious undertaking Professor McFarland computed, by means of both formulae, the eccentricity of the earth's orbit and the longitude of the perihelion for no fewer than 485 separate epochs. See "American Journal of Science," vol. xx. p. 105 (1880).

then fall as snow. But Professor Newcomb actually states that he cannot accept the conclusion that this would lead to more snow.

Influence of a Snow-covered Surface.—I have argued that this accumulation of snow would lower the summer temperature, and tend to prevent the disappearance of the snow, and have assigned three reasons for this conclusion:—

First.—Direct radiation. The snow, for physical reasons well known, will cool the air more rapidly than the sun's rays will heat it. This is shown from the fact that in Greenland a snow and ice-covered country, a thermometer exposed to the direct radiation of the sun has been observed to stand above 100°, while the air surrounding the instrument was actually 12° below the freezing point. Professor Newcomb and also Mr. Hill* regard the idea that this could in any way favour the accumulation of snow as absurd. They think that in fact it would have directly the opposite effect. They have perceived only one-half of the result. It is quite true, as they affirm, that the cooling of the air by the snow will not prevent the melting of the snow, but the reverse. There is, however, another and far more important result overlooked in their objection. If the snow and ice-covered surface keeps the temperature of the air, in summer, below the freezing-point, which it evidently does in Greenland and in the Antarctic continent, the moisture of the air will fall as snow and not as rain. No doubt this is the chief reason why in those regions, even in the middle of summer, rain seldom falls, the precipitation being almost always in the form of snow, although at that very season the direct heat of the sun is often as great as in India. Were the snow and icy

* "Geological Magazine" for January, 1880, p. 12.

mantle removed a snow-shower in summer would be as rare a phenomenon in those regions as it would be in the south of England.

Second.—'The rays which fall on snow and ice are to a great extent reflected back into space. But those that are not reflected, but absorbed, do not raise the temperature, for they disappear in the mechanical work of melting the ice.'

This reason is also regarded as absurd. The heat of the sun during the perihelion summer would, he says, suffice to melt the whole accumulation of winter snow in three or four days. "The reader," he continues, "can easily make a computation of the incredible reflecting power of the snow and of the unexampled transparency of the air required to keep the snow unmelted for three or four months." Incredible as it may appear to Prof. Newcomb, I shall shortly show that a less amount of snow than the equivalent of the two feet of ice which he assumes does actually, in some places, defy the melting-power of a tropical sun. But he misapprehends my reasoning here also, by overlooking the more important factor in the affair, namely, the keeping of the air in the summer below the freezing-point. The direct effect that this has in preventing the sun from melting the snow and ice will be discussed shortly; but the point to which I wish at present to direct special attention is the fact that if the air is kept below, or even at the freezing-point, snow will fall and not rain. Snow is a good reflector of heat; consequently a large portion of the sun's rays falling on the snow and icy surface is reflected back into space. The aqueous vapour of the air, on the other hand, as the vibrations of its molecules agree in *period* with those of the snow and ice, cuts off a large portion of the heat *radiated* by the snow surface; but here in the case of *reflection* under consideration the rays are not cut off; for the reflected rays are of the same character as the incident rays which pass so freely through the aqueous vapour. And in respect to the remaining rays which are not reflected, but absorbed by the snow, they do not manage to raise the temperature of the snow above the freezing-point. Consequently the air is kept in the condition most favourable for the production of snow.

Third.—'Snow and ice lower the temperature by chilling the air and condensing the vapour into thick fogs. The great strength of the sun's rays during summer, due to his nearness at that season, would, in the first place, tend to produce an increased amount of evaporation. But the presence of snow-clad mountains and an icy sea would chill the atmosphere and condense the vapour into thick fogs. The thick fogs and cloudy sky would effectually prevent the sun's rays from reaching the earth, and the snow in consequence would remain unmelted during the entire summer.'

On this Professor Newcomb's criticism is as follows: —"Here he (Mr. Croll) says nothing about the latent heat set free by the condensation, nor does he say where the heat goes to which the air must lose in order to be chilled. The task of arguing with a disputant who in one breath maintains that the transparency of the air is such that the rays reflected from the snow pass freely into space, and in the next breath that thick fogs effectually prevent the rays ever reaching the snow at all, is not free from embarrassment."

If he really supposes my meaning to be that the air is so transparent as to allow the incident and reflected rays of the sun to pass freely without interruption while at the same time and in the same place the air is not transparent but filled with dense fogs which effectually cut off the sun's rays and prevent them from reaching the earth, then I do not wonder that he should feel embarrassed in arguing with me. But if he supposes my meaning to be, as it of course is, that those two opposite conditions, existing at totally different times or in totally different places at the same time, should lead to similar results, namely the cooling of the air and consequent conservation of snow, then there is no ground whatever for any embarrassment about the matter.

"We might therefore show," he states, "that if the snow, air, fog, or whatever throws back the rays of the sun into space is so excellent a reflector of heat, it is a correspondingly poor radiator; and the same fog which will not be dissipated by the summer heat will not be affected by the winter's cold, and will therefore serve as a screen to prevent the radiation of heat from the earth during the winter."

There are few points in connection with terrestrial physics which appear to be so much misunderstood as that of the influence of fogs on climate. One chief cause of these misapprehensions is the somewhat complex nature of the subject arising from the fact that aqueous vapour acts so very differently under different conditions. When the vapour exists in the air as an invisible gas, we have often an intensely clear and transparent sky, allowing the sun's rays to pass to the ground with little or no interruption; and if the surface of the ground be covered with snow, a large portion of the incident rays are reflected back into space without heating either the snow or the air. The general effect of this loss of heat is, of course, to lower the general temperature. But when this vapour condenses into thick fogs it acts in a totally different manner. The transparency to a great extent disappears, and the fog then cuts off the sun's rays and prevents them from reaching the ground. This it does in two different ways. 1st. Its watery particles, like the crystals of the snow, are good reflectors, and the upper surface of the mass of fog on which the rays fall acts as a reflector, throwing back a large portion of the rays into stellar space. The rest of the rays which are not reflected enter the fog and the larger portion of them are absorbed by it. But it will be observed that by far the greater part of the absorption, if not nearly all of it, will take place in the upper half of the mass. This is a necessary result of a recognised principle in radiant heat known as the "sifting" of the rays. The deeper the rays penetrate into the fog, the less will be the amount of heat absorbed. If the depth of the mass be great, absorption will probably entirely disappear before the surface of the ground is reached. The fog will begin, of course, to radiate off the heat thus absorbed; but as it is the upper half of the mass which has received the principal part of the heat, the most of this heat will be radiated upward into stellar space, and, like the reflected heat, entirely lost in so far as heating the earth is concerned. A portion will also be radiated downward, some of which may reach the ground, but the greater portion will be reabsorbed in its passage through the mass. We have no means of estimating the amount of heat which would thus be thrown off into space by reflection and radiation; but it is certainly great. I think we may safely conclude that in places like South Georgia and Sandwich Land, where fogs prevail to such an extent during summer, one-half at least of the heat from the sun never reaches the ground. A deprivation of sun-heat of a much less extent than this would certainly lower the summer temperature of these places far below the freezing point, were it not for a compensating cause to which I shall now refer, viz. the heat "trapped" by the fog. The fog, although it prevents a large portion of the sun's heat from ever reaching a place, at the same time prevents to a great extent that place from losing the little heat which it does receive. In other words, it acts as a screen preventing the loss of heat by radiation into space. But the heat thus "trapped" never fully compensates for that not received, and a lowering of temperature is always the result.

Had all those considerations been taken into account by Professor Newcomb, Mr. Hill, Mr. Searles Wood, and others, they would have seen that I had by no means over-estimated the powerful influence of fogs in lowering the summer temperature.

The influence of fogs on the summer temperature is a fact so well established by observation that it seems strange that anyone should be found arguing against it.

Heat Evolved by Freezing.—There is one objection to which I may here refer, and which has been urged by nearly all my critics. It is said, correctly enough, that as water in freezing evolves just as much heat as is required to melt it, there is on the whole no actual loss of heat; that whatever heat may be absorbed in the mechanical work of melting the snow, just as much was evolved in the formation of the snow. Consequently it is inferred, in so far as climate is concerned, the one effect completely counterbalances the other. This inference, sound as it may at first sight appear, has been so well proved to be incorrect by Mr. Wallace that I cannot do better than quote his words:—

"In the act of freezing, no doubt water gives up some of its heat to the surrounding air, but *that air still remains below the freezing-point*, or freezing would not take place. The heat liberated by freezing is therefore what may be termed low-grade heat—heat incapable of melting snow or ice; while the heat absorbed while ice or snow is melting is high-grade heat, such as is capable of melting snow and supporting vegetable growth. Moreover, the low-grade heat liberated in the formation of snow is usually liberated high up in the atmosphere, where it may be carried off by winds to more southern latitudes; while the heat absorbed in melting the surface of snow and ice is absorbed close to the earth, and is thus prevented from warming the lower atmosphere which is in contact with vegetation. The two phenomena, therefore, by no means counterbalance or counteract each other, as it is so constantly and superficially asserted that they do" ("Island Life," p. 140).

The Fundamental Misconception.—I come now to a misapprehension which more than any other has tended to prevent a proper understanding of the causes which lead to the conservation by snow. Whatever the eccentricity of the earth's orbit may be, the heat received from the sun during summer is more than sufficient to melt the snow of winter. Consequently, it is assumed no permanent accumulation of snow can take place. This objection, as expressed by Mr. Hill, is as follows: "We have no reason to suppose that at present, in the northern hemisphere, more snow or ice is anywhere formed in winter than is melted in summer. With greater eccentricity, less heat than now would be received in winter, but exactly as much more in summer. More snow would therefore be formed in the one half of the year, but exactly as much more be melted in the other half. The colder winter and the warmer summer would exactly neutralize each other's effects, and on the average of years no accumulation could begin. *Primâ facie,* therefore, high eccentricity will not account for glacial periods." * In the language of Prof. Newcomb, it is as follows:— "During this perihelion summer, the amount of heat received from the sun by every part of the northern hemisphere would suffice to melt from four to six inches of ice per day over its entire surface; that is, it would suffice to melt the whole probable accumulation in three or four days. The reader can easily make a computation of the incredible reflecting power of the snow and of the unexampled transparency of the air required to keep the snow unmelted for three or four months."

It is assumed in this objection that because the heat received from the sun by an area is more than sufficient to melt all the snow that falls on it, no permanent accumulation of snow and ice can take place. It is assumed that the quantity of snow and ice melted must be proportional to the heat received. Suppose that on a certain area a given amount of snow falls annually. The amount of heat received from the sun per annum is computed; and after the usual deduction for that cut off by the atmosphere has been made, if it be found that the quantity remaining is far more than sufficient to melt the snow, it is then assumed that the snow must be melted, and that no accumulation of snow and ice year by year in this area is possible. To one approaching this perplexing subject for the first time, such an assumption looks very plausible; but a little reflection will show that it is most superficial. The assumption is at the very outset totally opposed to known facts. Take the lofty peaks of the Himalayas and Andes as an example. Few, I suppose, would admit that at these great elevations as

* "Geological Magazine," January, 1880, p. 12.

much as 50 per cent. of the sun's heat could be cut off. But if 50 per cent. reaches the snow, this would be sufficient to melt fifty feet of ice; and this, no doubt, is more than ten times the quantity which actually requires to be melted. Notwithstanding all this, the snow is never melted, but remains permanent. Take, as another example, South Georgia, in the latitude of England. Suppose we assume that one half of the sun's heat is cut off by the clouds and fogs which prevail to such an extent in that place, still the remaining half would be sufficient to melt upwards of thirty feet of ice, which is certainly more than the equivalent of all the snow which falls; yet this island is covered with snow and ice down almost to the sea-shore during the whole year. Take still another example, that of Greenland. The quantity of heat received between latitudes 60° and 80°, which is that of Greenland, is, according to Meech, one half that received at the equator; and were none cut off, it would be sufficient to melt fifty feet of ice. The annual precipitation on Greenland in the form of snow and rain, according to Dr. Rink, amounts to only twelve inches; and two inches of this he considers is never melted, but is carried away in the form of ice-bergs. Mr. Hill maintains* that, owing to the great thickness of the air traversed by the sun's rays, and the loss resulting from the great obliquity of reflection, the amount of heat reaching the ground would be insufficient to melt more than sixteen feet of ice. Supposing we admit this estimate to be correct, still this is nineteen times more than is actually melted. The sun melts only ten inches, notwithstanding the fact that it has the power to melt sixteen feet.

* "Geological Magazine," April, 1880.

In short, there is not a place on the face of the globe where the amount of heat received from the sun is not far more than sufficient to melt all the snow which falls upon it. If it were true, as the objection assumes, that the amount of snow melted is proportional to the amount of heat received by the snow, then there could be no such thing as perpetual snow.

The reason why the amount of snow and ice melted is not necessarily proportional to the amount of heat received is not far to seek. Before snow or ice will melt, its temperature must be raised to the melting-point. No amount of heat, however great, will induce melting to begin unless the intensity of the heat be sufficient to raise the temperature to the melting-point. Keep the temperature of the snow below that point, and, though the sun may shine upon it for countless ages, it will still remain unmelted. It is easy to understand how the snow on the lofty summits of the Himalayas and the Andes never melts. According to the observations made at Mount Whitney, to which reference has already been made, the heat of even a vertical sun would not be sufficient at these altitudes to raise the temperature of the snow to near the melting-point; and thus melting, under these conditions, is impossible. The snow will evaporate, but it cannot melt. But, owing to the frozen condition of the snow, even evaporation will take place with extreme difficulty. If the sun could manage to soften the snow-crystals and bring them into a semi-fluid condition, evaporation would, no doubt, go on rapidly ; but this the rays of the sun are unable to do; consequently we have only the evaporation of a solid, which, of course, is necessarily small.

It may here be observed that at low elevations, where the snowfall is probably greater, and the amount of heat received even less, than at the summits, the snow melts and disappears. Here, again, the influence of that potent agent, aqueous vapour, comes into play. At high elevations the air is dry, and allows the heat radiated from the snow to pass into space; but at low elevations a very considerable amount of the heat radiated from the snow is absorbed by the aqueous vapour which it encounters in passing through the atmosphere. A considerable portion of the heat thus absorbed by the vapour is radiated back on the snow; but the heat thus radiated being of the same quality as that which the snow itself radiates, is on this account absorbed by the snow. Little or none of it is reflected, like that received from the sun. The consequence is, that the heat thus absorbed accumulates in the snow till melting takes place. Were the amount of aqueous vapour possessed by the atmosphere sufficiently diminished, perpetual snow would cover our globe down to the sea-shore. It is true that the air is warmer at the lower than at the higher levels, and, by contact with the snow, must tend to melt it more at the former than at the latter position. But we must remember that the air is warmer mainly in consequence of the influence of aqueous vapour, and that, were the quantity of vapour reduced to the amount in question, the difference of temperature at the two positions would not be great.

But it may be urged, as a further objection to the foregoing conclusion, that, as a matter of fact, on great mountain-chains the snow-line reaches to a lower level on the side where the air is moist than on the opposite side where it is dry and arid—as, for example, on the southern side of the Himalayas and on the eastern side of the Andes, where the snow-line descends 2000 or 3000 feet below that of the opposite or dry side.

But this is owing to the fact that it is on the moist side that by far the greatest amount of snow is precipitated. The moist winds of the south-west monsoon deposit their snow almost wholly on the southern side of the Himalayas, and the south-east trades on the east side of the Andes. Were the conditions in every respect the same on both sides of these mountain-ranges, with the exception only that the air on one side was perfectly dry, allowing radiation from the snow to pass without interruption into stellar space, while on the other side the air was moist and full of aqueous vapour absorbing the heat radiated from the snow, the snow-line would in this case undoubtedly descend to a lower level on the dry than on the moist side. Melting would certainly take place at a greater elevation on the moist than on the dry side; and this is what would mainly determine the position of the snow-line.

The annual precipitation on Greenland, as we have seen, is very small, scarcely one-half that of the driest parts of Great Britain. This region is covered with snow and ice, not because the quantity of snow falling on it is great, but because the quantity melted is small; and the reason why the snow does not melt is not that the amount of heat received during the year is unequal to the work of melting the ice, but that, mainly through the dryness of the air, the snow is prevented from rising to the melting-point. The very cause which prevents a heavy snowfall protects the little which does fall from disappearing. The same remarks apply to the Antarctic regions.

In South Georgia and Fuego, where clouds and dense fogs prevail during nearly the whole year, the permanent snow and ice are due to a different cause. Here the snowfall is great, and the amount of heat cut off enormous; but this alone would not account for the non-disappearance of the snow and ice; for, notwithstanding this, the heat received is certainly more than sufficient to melt all the snow which falls, great as that amount may be. The real cause is that the heat received is not sufficiently intense to raise the temperature to the melting-point. More heat is actually received by the snow than is required to melt it; but it is dissipated and lost before it can manage to raise the temperature of the snow to the melting-point; consequently the snow is not melted. Here snow falls in the very middle of summer; but snow would not fall unless the temperature were near the freezing-point.

Foregoing principles applied to the case of the Glacial Epoch.—Let us now apply the foregoing principles to the case of the glacial epoch. As winter then occurred in aphelion during a high state of eccentricity, that season would be much longer and colder than at present. Snow in temperate regions would then fall in place of rain; and although the snowfall during the winter might not be great, yet, as the temperature would be far below the freezing-point, what fell would not melt. As heat, which produces *evaporation,* is just as essential to the accumulation of snow and ice as is cold, which produces *condensation,* after the sun had passed the vernal equinox and summer was approaching, the consequent rise of temperature would be accompanied by an increase in the snowfall. A melting of the snow would also begin; but it would be a very considerable time before the amount melted would equal the daily amount of snow falling. Rain, alternating with snow-showers, would probably result; and, for some time before midsummer, snow would cease and give place entirely to rain. Melting would then go on rapidly, and by the end of the summer the snow would all disappear except on high mountain-summits such as those of Scotland, Wales, and Scandinavia. Before the end of autumn, however, it would again begin to fall. Next year would bring a repetition of the same process, with this difference, however, that the snow-line would descend to a lower level than in the previous year. Year by year the snow-line would continue to descend till all the high grounds became covered with permanent snow.

It would not require a very great amount of change from the present condition of things to bring about such a result. A simple lowering of the temperature, which would secure that snow, instead of rain, should fall for six or eight months in the year, would suffice; and this would follow as a necessary result from an increase of eccentricity. Now, if all our mountain-summits were covered with permanent snow down to a considerable distance, the valleys would soon become filled with local glaciers. In such a case we should then have more than one-half of Scotland, a large part of the north of England and Wales, with nearly the whole of Norway, covered with snow and ice. Here a new and powerful agent would come into operation which would greatly hasten on a glacial condition of things. This large snow-and-ice-covered surface would tend to condense the vapour into snow. It would, during summer, chill the air and produce dense and continued fogs, cutting off the sun's rays, and leading to a state of things approaching to that of South Georgia, which would much retard the melting of the snow.

It is a great mistake, as I have repeatedly shown, to suppose that the perihelion summers of the glacial epoch could be hot. No snow-and-ice-covered continent can enjoy a hot summer. This is clearly shown by the present condition of Greenland. Were it not for the ice, the summers of North Greenland, owing to the continuance of the sun above the horizon, would be as warm as those of England; but, instead of this, the Greenland summers are colder than our winters, and snow during that season falls more or less nine days out of ten. But were the ice-covering removed, a snow-shower during summer would be as great a rarity as it would be with us. On the other hand, were India covered with an ice-sheet, the summers of that place would be colder than those of England.

When the high grounds of Scotland and Scandinavia, with those of the northern parts of America, became covered with snow and ice, and the eccentricity went on increasing, a diminution of the Gulf Stream, and a host of other physical agencies, all tending towards a glacial condition of things, would be brought into operation. This would ultimately and inevitably lead to a general state of glaciation, without the aid of any of those *additional* geographical changes of land and water which some have supposed.

The Mutual Reaction of the Physical Agents.—Those who think that the agencies to which I refer would not by themselves bring about a glacial condition appear to overlook a most important and remarkable circumstance regarding their mode of operation, to which I have frequently alluded in 'Climate and Time' (pp. 74-77) and other places. The circumstance is this:—The physical agencies in question not only all lead to one result, viz. an accumulation of snow and ice, but their efficiency in bringing about this result is actually strengthened by their mutual reaction on one another. In physics the effect reacts on the cause. In electricity and magnetism, for example, cause and effect in almost every case mutually act and react upon each other; but the reaction of the effect tends to weaken the cause. Those physical agents to which I have referred, no doubt, in their mutual actions and reactions, obey the same law; but in reference to *one particular result,* viz. the accumulation and conservation of snow, those mutual reactions strengthen one another. This is not reasoning in a circle, as Mr. Searles Wood supposes; for the reaction of an effect may on the whole weaken the cause, and yet in regard to a particular result it may strengthen it. In the case under consideration the agents not only act in one direction, but their efficiency in acting in that one direction is strengthened by their mutual reactions. This curious circumstance throws a flood of light on the causes which tended to bring about the glacial epoch.

To begin with, we have a high state of eccentricity. This leads to long and cold winters. The cold leads to snow; and although heat is given out in the formation of the snow, yet the final result is that the snow intensifies the cold: it cools the air and leads to still more snow. The cold and snow bring a third agent into play—*fogs,* which act still in the same direction. The fogs intercept the sun's rays; this interception of the rays diminishes the melting-power of the sun, and so increases the accumulation. As the snow and ice continue to accumulate, more and more of the rays are cut off; and, on the other hand, as the rays continue to be cut off, the *rate* of accumulation increases, because the quantity of snow and ice melted becomes thus annually less and less. In addition, the loss of the rays cut off by the fogs lowers the temperature of the air and leads to more snow being formed, while, again, the snow thus formed chills the air still more and increases the fogs. Again, during the winters of a glacial epoch, the earth would be radiating its heat into space. Had this loss of heat simply lowered the temperature, the lowering of the temperature would have tended to diminish the rate of loss; but the result is the formation of snow rather than the lowering of the temperature.

Further, as snow and ice accumulate on the one hemisphere they diminish on the other. This increases the strength of the trade-winds on the cold hemisphere and weakens those on the warm. The effect of this is to impel the warm water of the tropics more to the warm hemisphere than to the cold. Suppose the northern hemisphere to be the cold one; then, as the snow and ice begin gradually to accumulate, the ocean currents of that hemisphere, more particularly the Gulf Stream, begin to decrease in volume, while those on the southern or warm hemisphere begin *pari passu* to increase.* This withdrawal of heat from the northern hemisphere favours the accumulation of snow and ice; and as the snow and ice accumulate the ocean currents decrease. On the other hand, as the ocean currents diminish the snow and ice still more accumulate. Thus the two effects, in so far

* Prof. Dana has shown that in North America those areas which at present have the greatest rainfall are, as a rule, the areas which were most glaciated during the glacial epoch. Mr. Searles V. Wood ("Geol. Mag.," July, 1883) maintains that this fact is inconsistent with the theory that the glacial period was due to the cause to which I attribute it. I am totally unable to comprehend how he arrives at this conclusion. Supposing the Gulf Stream, as I have maintained, were greatly diminished during the glacial period, still I think it would follow, other things being equal, that the areas which now have the greatest rainfall would during that period probably have the greatest snowfall, and consequently the greatest accumulation of ice. The amount of precipitation might be less than at present; but this would not prevent the areas which had the greatest snowfall from being most covered with ice.

as the accumulation of snow and ice is concerned, mutually strengthen each other.

The same process of mutual action and reaction takes place among the agencies in operation on the warm hemisphere; only the result produced is diametrically opposite to that produced in the cold hemisphere. On this warm hemisphere action and reaction tend to raise the mean temperature and diminish the quantity of snow and ice existing in temperate and polar regions.

The primary cause of all these physical agencies being set in operation is a high state of eccentricity of the earth's orbit; and with a continuance of that state a glacial epoch becomes inevitable.

The Explanation begins with Winter.—Mr. Hill asks why I always begin in my explanation with the aphelion winter rather than with the perihelion summer. The reason is that the character of the summer is determined by that of the winter, and not the winter by that of the summer. It is true that, to a certain extent, the influence is mutual; but the effect of the summer on the winter is trifling in comparison with that of the winter on the summer. To begin our explanation with the summer would be like beginning at the end of a story and telling it backward.

M. Woeikof on the Cause of Glaciation.—In an article by A. Woeikof on "Glaciers and Glacial Periods in their Relations to Climate" ("Nature," March 2nd, 1882), it is maintained that the chief cause which leads to the formation of snow, and consequently to a glacial condition, is a low surface-temperature of the sea surrounding or adjoining the land. When the surface-temperature of the water much exceeds the freezing-point, the vapour, he says, evaporated from the sea and condensed on the land will be rain and not snow;

but when the temperature of the water is near the freezing-point, snow will be the result. A diminution, for example, in the heat brought by the Gulf Stream that would very greatly lower the surface-temperature of the sea surrounding Great Britain would, he says, bring about a heavy snowfall, and lead to permanent snow and ice. Again, he maintains, "As there is no reason to suppose that the surface-temperature of the sea would be lower during winter in aphelion and high eccentricity, it follows that there will not be more snow than now in countries where rain is the rule, even in winter, all other things equal."

There is surely a fallacy lurking under this theory of M. Woeikof. Snow instead of rain is not, as he supposes, owing to the low temperature of the water from which the vapour is derived, but to the low temperature of the air where the vapour is precipitated. Of course, when the surface of the sea is near the freezing-point, the air over the sea and the adjoining land is usually also not far from the freezing-point, and consequently the precipitation is more likely to be snow than rain. If the air be cold, as it generally is over a snow- and ice-covered country, a high temperature of the adjoining seas, were this possible, would greatly increase the snowfall, because it would greatly augment the quantity of vapour which would be available for snow.

CHAPTER IV.

OBJECTION THAT THE AIR AT THE EQUATOR IS NOT HOTTER IN JANUARY THAN IN JULY.

Influence of the present distribution of Land and Water.—The Summer of the Southern Hemisphere colder than that of the Northern.—Influence of the Trade Winds on the Temperature of the Equator.

THE fact that the temperature of the equator in January, when the earth is in perihelion, is not higher than in July, when in aphelion, has been urged as an objection to the Physical Theory. When we examine, however, the reason of this apparently curious circumstance, we find that there is certainly no real grounds for the objection.

The temperature referred to is, of course, the ordinary temperature as indicated by the shade thermometer, which is simply that of the air. The objection is more apparent than real; for if we examine the *indirect* results which follow from the present distribution of land and water, we shall see that there is no reason whatever why the air at the equator should be hotter in January than in July.

It is well known that, notwithstanding the nearness of the sun in January, the influence of the present distribution of land and water is sufficient to make the mean temperature of the whole earth, or, what is the same, the mean temperature of the air over the surface of the earth higher in July than in January.

The reason of this is obvious. Nearly all the land is in the northern hemisphere, while the southern hemisphere is for the most part water. The surface of the northern or land-hemisphere, for reasons which have been discussed in the last chapter, becomes heated in summer and cooled in winter to a far greater extent than the surface of the southern or water hemisphere. Consequently, when we add the July or midsummer temperature of the northern to the July temperature of the southern hemisphere, we must get a higher number than when we add the January or midwinter temperature of the former to the January temperature of the latter. For example, the mean July temperature of the northern hemisphere, according to Dove ("Distribution of Heat on the Surface of the Globe") is 70°·9, and that of the southern hemisphere 53°·6; add the two together and we have 124°·5, which gives a mean for both hemispheres of 62°·3. The mean January temperature of the northern hemisphere is 48°·9, which, added to 59°·5, the mean January temperature of the southern hemisphere, gives only 108°·4, or a mean of 54°·2. Consequently the air over the surface of the globe is hotter in July by 8° than in January, notwithstanding the effects of eccentricity. It is obvious that, were it not for the counteracting effects of eccentricity, the difference would be much greater. Ten thousand years ago, when eccentricity and the distribution of land and water combined to produce the same effect, the difference must have been far greater than 8°.

But it will be asked, How can this affect the air over the equator, which is not situated more on the one hemisphere than on the other? It is true that those causes have but little *direct* effect on the air at the equator, but *indirectly* they have a very powerful influence. The air is continually flowing in to the equatorial regions from both hemispheres. In fact, the air which we find there is derived entirely from the temperate regions. In July we have the northern trades coming from a hemisphere with a mean temperature as high as 70°·9, and the southern trades coming from a hemisphere with a mean temperature not under 53°, while in January the former trades flow from a hemisphere as low as 50°, and the latter from a hemisphere no higher than 60°. Consequently, the air which the equatorial regions received from the trades must have a higher temperature in July than in January. The northern is the dominant hemisphere; it pours in hot air in July and cold air in January, and this effect is not counterbalanced by the air from the opposite hemisphere. The mean temperature of the air passing into the equatorial regions ought, therefore, to be much higher in July than in January, and this it no doubt would be were it not, let it be observed, for the counteracting effects of eccentricity. The tendency of the present distribution of land and water, when our northern winter occurs in perihelion, is to counteract the effects of eccentricity. But ten thousand years ago, when our winters were in aphelion, that cause would co-operate to intensify the effects of eccentricity. In fact, it would actually more than double the effects then produced by eccentricity. Now if the influence of the present distribution of land and water is so great as not merely to counteract but to reverse the effects of eccentricity to the extent of making the mean temperature of the earth 8° warmer in July than in January, it is not surprising that it should be sufficient to make the equatorial regions at least as warm in the former as in the latter period.

The fact that the equator at present is not hotter when the earth is in perihelion, instead of being an objection to the theory that the glacial period was due to an increase of eccentricity, as is supposed by some, is in reality another strong argument in its favour, for it shows that a much less amount of eccentricity would suffice to induce a commencement of glacial conditions in the northern hemisphere than would otherwise be required, were it not for the circumstances to which reference has been made. This objection, like many others which have been urged against the theory, arises from looking too exclusively at the *direct* effects of eccentricity.

There is another cause which must also tend to lower the January and raise the July temperature of the equator, viz., the northern trades pass farther south in January than in July, and consequently cool the equatorial regions more during the former than the latter season. This general tendency of the trades to lower the temperature of the equatorial regions more in January than in July is, of course, subject to modifications from the monsoons, the rainy seasons, and other local causes; nevertheless, so long as the present distribution of land and water endures, so long will eccentricity have a counteracting effect upon the temperature of the air at the equator, which but for that would be hotter in July than in January.

No knowledge whatever as to the intensity of the sun's heat can be obtained from observations on the temperature of the air at the equator. The comparatively cold air flowing in from the temperate regions has not time to be fully heated by the sun's rays before it rises as an ascending current and returns to the temperate regions from whence it came. More than this, these trades prevent us from being able to determine with accuracy the intensity of the sun's heat from the temperature of the ground; for the surface of the ground in equatorial regions is kept at a much lower temperature by the air blowing over it than is due to the intensity of the sun's heat. It thus becomes a very intricate problem to determine how much the surface of the ground is kept below the maximum temperature by the heat absorbed by the moving air.

CHAPTER V.

THE ICE OF GREENLAND AND THE ANTARCTIC CONTINENT NOT DUE TO ELEVATION OF THE LAND.

Greenland; attempts to Penetrate into the Interior.—No Mountain Ranges in the Interior.—The Föhn of Greenland.—Antarctic Regions.—Character of the Icebergs.—Sir Joseph Dalton Hooker and Professor Shaler on the Antarctic Ice.—On the Argument against the Existence of a South-Polar Ice-Cap.—Thickness of Ice not dependent on Amount of Snowfall.

BEFORE proceeding to an examination of certain modifications of the Physical Theory which have recently been advanced, it will be necessary to devote the present chapter to the consideration of some points connected with the physical conditions of the ice of Greenland and the Antarctic regions.

The only two continents on the globe covered by permanent ice and snow are Greenland and the Antarctic. But are these continents to be regarded as Highlands or as Lowlands? It is an opinion held by many that these regions are greatly elevated, and that it is mainly owing to this elevation that they are so completely buried under ice. I have been wholly unable to find evidence for any such conclusion. It is of course true that, in regard to Greenland at least, the observations of Rink, Heyes, Nordenskjöld, Jensen, Brown, and others, show that the upper surface of the inland ice is greatly elevated above the sea-level. Dr. Rink, for example, states that the elevation of this icy plain, at its junction with the outskirts of the country where it begins to lower itself through the valleys, in the ramifications of the Bay of Omenak is about 2000 feet, from which it gradually *rises towards the interior*. Nordenskjöld, in his first journey on the inland ice, 30 miles from the coast, reached an elevation of 2200 feet, and found the ice continued to *rise inwards*. Hayes, who penetrated 50 miles into the interior, found the elevation about 5000 feet, and still *continuing to slope upwards towards the interior* of the continent.

The mystery of the interior of Greenland has at last been cleared up by Baron Nordenskjöld in his recent expedition. It was found, as might have been expected, that the interior of Greenland is a complete desert of ice, with the icy plain gradually sloping upwards towards the ice-shed or centre of dispersion. After penetrating to a distance of 280 miles from the coast, the surface of the icy plain was found to be no less than 7000 feet above the sea-level; and this plain was still seen to rise to the east. The greater part of the surface of the inland ice is, of course, far above the snow-line; but this by no means proves that Greenland is an elevated country, for this elevation of the upper surface of the ice may be due *entirely* to the thickness of the sheet.

Of all the results gained by Nordenskjöld's famous expedition, perhaps the most important is the confirmation it has afforded of the true nature of continental ice.

Certainly no one has ever seen, and probably no one ever will see, elevated land under the ice either of Greenland or the Antarctic continent; and to assume its existence because those regions are so completely glaciated would simply be to beg the very question at issue.

It will doubtless be urged that, although the ground under the ice may not be elevated, yet there may be lofty mountain-chains in the interior which might account for the origin of the ice. We have, I think, good grounds for concluding that if there are mountain-ranges in the interior of Greenland (of which there is absolutely no proof, although one or two isolated peaks have been seen), they must be wholly buried under the ice. For, if mountain-masses rise above the icy mantle, there ought to be evidence of this in the form of broken rock, stones, earth, and other moraine matter lying on the inland ice. "But as soon as we leave the immediate vicinity of the coast," says Dr. Brown, "no moraine is seen coming over the inland ice: no living creature, animal or plant, except a minute alga." And Baron Nordenskjöld in his recent expedition over the inland ice says, "After a journey of about half a kilometre from the ice border no stone was found on the surface, not even one as large as a pin's point." This could not possibly be the case if ranges of mountains rose above the general ice-covering. These mountain-ranges, if they exist, are doubtless covered with snow, and their sides with glaciers; but this would not prevent pieces of broken rock and stones from rolling down upon the inland ice. In fact, it would have the very opposite result; for glaciers would be one of the most effective agents possible in bringing down such material, and it is certain that no avalanche of snow could take place without carrying along with it masses of stones and rubbish. All these materials brought down from the sides of the projecting peaks would be deposited on the surface of the inland ice and carried along with it in its outward motion from the centre of dispersion, and could not fail to be observed did they exist. The fact that no such thing is ever seen is conclusive proof that these supposed projecting mountain-ranges do not exist.

But it may still be urged that the absence of moraine matter on the surface of the inland ice is not sufficient evidence that they do not exist; for as this material from the interior would have to travel hundreds of miles before reaching the outskirts, a journey occupying a period of many years, the stones would become buried under the successive layers of ice formed on the surface during their passage outwards. But supposing this were the case, these buried moraines, if they existed, ought to be seen projecting from the edge of the sheet at places where icebergs break off, and also on the edges of the icebergs themselves near their tops; but such, I presume, is never the case. Further, as the inland ice has to force its way through the comparatively narrow fjords before reaching the sea, the moraines could not fail to be occasionally observed did they exist.

But supposing there were mountains in the interior, this would not account for the general ice-covering. It would not account for the intervening spaces between the mountains being filled up with ice. To account for the whole country being covered with ice through the influence of mountains, we should have to assume that it was studded over with them at no great distance from one another; otherwise, all that we should have would simply be local glaciers.

Dr. Robert Brown, one of the highest authorities in matters relating to Greenland, who does not believe in the existence of mountain-masses in the interior, says: —"I do not think a range of mountains at all necessary for the formation of this huge *mer de glace*, for this idea is derived from the Alpine and other mountain ranges, where the glacial system is a petty affair compared with that of Greenland. I look upon Greenland," he continues, "and its interior ice-field in the light of a broad-lipped shallow vessel, but with breaks in the lip here and there, and the glacier like some viscous matter in it. As more is poured in, the viscous matter will run over the edges, naturally taking the line of the chinks as its line of outflow. The broad lips of the vessel, in my homely *simile*, are the outlying islands or 'outskirts'; the viscous matter in the vessel, the inland ice; the additional matter continually being poured in, the enormous snow-covering, which, winter after winter, for seven or eight months in the year, falls almost continuously on it; and the chinks or breaks in the vessel are the fjords or valleys down which the glaciers, representing the outflowing viscous matter, empty the surplus of the vessel."*

In North Greenland and along Smith Sound a warm south-east wind, somewhat similar to the *Föhn* of Switzerland, has been reported in the middle of winter. From this it has been inferred by some that there must be high ranges of mountains in the interior from which this wind descends. There are, however, certainly no good grounds for such a conclusion; for we know that the upper surface of the inland ice of North Greenland, 50 or 100 miles from the outskirts, has an elevation of at least 4000 or 5000 feet. Now, a wind crossing this icy plateau and descending to the sea level would have its temperature raised by upwards of 20°, and also its capacity for moisture at the same time greatly increased. The consequence would therefore be that, in the midst of a Greenland winter, such a wind would be felt to be hot and dry.

The opinion was expressed by Giesecke, who long resided in Greenland, that that country is merely a

* "Arctic Papers for the Expedition of 1875," p. 24.

collection of islands fused together by ice. This opinion is concurred in by Dr. Brown, who says that "most likely it will be found that Greenland will end in a broken series of islands forming a Polar archipelago. That the continent (?) is itself a series of such islands and islets—consolidated by means of the inland ice—I have already shown to be highly probable, if not absolutely certain, as Giesecke and Scoresby affirmed." It has long been a belief that several of the west-coast fjords cut through Greenland from sea to sea—in short, that they are simply straits filled up with ice. The important bearing that this island-condition of Greenland has on the explanation of the warm interglacial periods of that country will be shown in a future chapter.

Antarctic Regions.—It need hardly be remarked, that what has been stated as to the total absence of proof that Greenland possesses elevated plateaus and ranges of lofty mountains holds in a still more marked degree in reference to the Antarctic continent. Here is a region nearly 3000 miles across, buried under ice, on which the foot of man never trod. There is not the shadow of a basis for concluding that the interior of this immense region is, under the ice, greatly elevated, or that it possesses lofty mountain-ranges. The probability seems rather to be that, like Greenland, the area, as Sir Wyville Thomson supposes, consists of comparatively low dismembered land or groups of islands bound together by a continuous sheet of ice. "We have no evidence," says Sir Wyville, "that this space, which includes an area of about 4,500,000 square miles, nearly double that of Australia, is continuous land. The presumption would seem rather to be that it is at all events greatly broken up; a large portion of it probably consisting of groups of low islands united and combined by an extension of the ice-sheet."

"Various patches of Antarctic land," he continues, "are now known with certainty, most of them between the parallels of 65° and 70° S.; most of these are comparatively low, their height, including the thickness of their ice-covering, rarely exceeding 2000 to 3000 feet. The exceptions to this rule are Victoria Land and the volcanic chain, stretching from Balleny Island to latitude 78° S.; and a group of land between 55° and 95° west longitude, including Peter the Great Island, Alexander Island, Graham Land, Adelaide Island, and Louis Philippe Land. The remaining Antarctic land, including Adelie Land, Clairie Land, Sabrina Land, Kemp Land, and Enderby Land, nowhere rises to any great height."*

There is another class of facts which shows still more conclusively the probably low flat nature of most of the Antarctic regions. I refer to the character of the great ice-barrier, and the bergs which break off from it. The icebergs of the southern ocean are almost all of the tabular form, and their surface is perfectly level, and parallel with the surface of the sea. The icebergs are all stratified; the stratifications running parallel with the surface of the berg. The stratified beds, as we may call them, are separated from each other by a well-marked blue band. These blue lines or bands, as Sir Wyville Thomson remarks, are the sections of sheets of clear ice; while the white intervening spaces between them are the sections of layers of ice where the particles are not in such close contact and probably contain some air. The blue bands, as Sir Wyville suggests, probably represent portions of the snow surface which

* "Lecture on Antarctic Regions" (Collins, Glasgow, 1877); "Nature," vol. xv.

during the heat of summer becomes partially melted and refrozen into compact ice; while the intervening white portions represent the snow of the greater part of the year, which of course would become converted into ice without ever being actually melted. It is, therefore, more than probable that each bed with its corresponding blue band may represent the formation of one year. Judging from the number of these layers in an iceberg, some of these bergs must be of immense age, occupying a period probably of several thousand years in their formation. And as the ice is in a constant state of motion outwards from the centre of dispersion—probably the South Pole—the bergs before becoming detached from their parent mass must have traversed a distance of hundreds of miles.

The fact that these bergs must have travelled from great distances in the interior is further evident from the following consideration. The distance between the well-marked blue lines is greatest near the top of the berg, where it may be a foot or more, and becomes less and less as we descend, until, near the surface of the water, it is not more than two or three inches. This diminution in the thickness of the ice-strata from the top downwards has been considered by Sir Wyville to be mainly due to two causes—compression, and melting of the ice, particularly the latter. But in my paper on the Antarctic Ice ("Quart. Journ. of Science," Jan. 1879) I have shown that, although compression and melting may have had something to do in the matter, this thinning of the strata from the top downwards is a necessary physical consequence of continental ice radiating from a centre of dispersion. Assuming the South Pole to be this centre, a layer which in, say, latitude 85° covers 1 square foot of surface will, on reaching latitude 80°, cover 2 square feet; at latitude

70° it will occupy 4 square feet, and at latitude 60° the space covered will be 6 square feet. Then, if the layer was 1 foot thick at latitude 85°, it would be only 6 inches thick at latitude 80°, 3 inches thick at latitude 70°, and 2 inches at latitude 60°. Had the square foot of ice come from latitude 89° it would occupy 30 square feet by the time it reached latitude 60°, and its thickness would be reduced to $\frac{1}{30}$ of a foot, or $\frac{2}{5}$ of an inch.

Now, the lower the layer the older it is, and the greater the distance which it has travelled. A layer near the bottom may have been travelling from the Pole for the past 10,000 or 15,000 years, whereas a layer near the top may perhaps not be 20 years old, and may not have travelled the distance of a mile. The ice at the bottom of a berg may have come from near the Pole, whereas the ice at the top may not have travelled 100 yards.

There is still another consideration which must be taken into account. It is this: the icebergs all seem to bear the mark of their original structure, and the horizontal stratifications appear also never to have been materially altered in their passage from the interior. This fact seems to have struck Sir Wyville forcibly. "I never saw," he says, "a single instance of deviation from the horizontal and symmetrical stratification which could in any way be referred to original structure, which could not, in fact, be at once accounted for by changes which we had an opportunity of observing taking place in the icebergs. There was not, so far as we could see, in any iceberg the slightest trace of structure stamped upon the ice in passing down a valley, or during its passage over *roches moutonnées* or any other form of uneven land. The only structure, except the parallel stratifications, which we ever observed which could be regarded as bearing upon the mode of original formation of the ice-mass, was an occasional local thinning-out of some of the layers and thickening of others—just such an appearance as might be expected to result from the occasional drifting of large beds of snow before they have time to become consolidated."

There cannot, I think, be the shadow of a doubt that these thin horizontal bands of clear blue ice, with their less dense and white intervening beds, are the original structure of the bergs. And it is evident that, if the ice had crossed mountain-ridges, valleys, or other obstructions in the course of its journey from the interior, these beds could not have avoided being crushed, fractured, broken up, and mixed together. Had this happened, it would have been physically impossible that they could ever have been restored to their old positions. Ice is, no doubt, plastic, and pressure, along with motion, might perhaps induce fresh lines of stratification; but neither motion nor pressure could have selected broken blue bands from among the white and placed them in their old positions.

Why the icebergs from Greenland are not of the tabular form, and stratified like those of the Antarctic regions, is doubtless owing to the fact that the Greenland ice is discharged through narrow fjords, which completely destroy the original horizontal stratifications.

Let us now see the consequence to which the foregoing considerations all lead. The tabular form and flat-topped character of the icebergs, with their perfectly horizontal bedding, show that they have been formed on a flat and even surface. They show also that this flat surface is not a mere local condition, but that it must be the general character of the Antarctic land; for all, or nearly all, of the bergs are of this tabular form. Again, the unaltered character of the stratifications of the bergs shows that there can be no great mountain-ranges, or even much rough and uneven ground, in the interior; for if there were, the bergs in their passage outwards would have had to pass over it; and this they could not have done and still have retained, as they actually have, their horizontal stratification undisturbed. These icebergs, as we have seen, must have traversed in their outward motion, before being disconnected with the ice-sheet, a distance of hundreds of miles; yet none of them bears the marks of having passed down or across a valley, or even over *roches moutonnées*.

That the Antarctic continent has a flat and even surface, the character of the icebergs shows beyond dispute. But this, it will be urged, does not *prove* that this surface may not be greatly elevated; in other words, that it may not be a flat elevated plateau. This, of course, is true; but it is evidently far more likely that this region, nearly 3000 miles across, should consist of flat, dismembered land, or groups of low islands separated and surrounded by shallow seas, than that it should consist of a lofty plateau without either hills, valleys, or mountain-ridges. In this case it may be that the greater part of the Antarctic ice-cap rests on land actually below sea-level, viz., on the floor of the shallow seas surrounding those island-groups. We know that such a condition of things was actually the case in regard to the great ice-sheet of North-western Europe during the glacial epoch. A glance at the Chart of the path of the ice given in 'Climate and Time,' p. 448 (and which is also reproduced in Chapter VIII. of the present volume), will show that the larger portion of the sheet rested on the bed of the Baltic, German Ocean, and the seas around Great Britain and Ireland and the Orkney and Shetland islands. That the Antarctic ice was formed on low and flat land, bordered for considerable distances by shoal water, was the opinion also of Sir Wyville Thomson.

Sir Joseph Dalton Hooker thinks that much of the Antarctic ice-sheet, thousands of feet in thickness as it is, was formed by the successive accumulations of snow year by year on pack-ice. The snowfall in the Antarctic regions he believes to be enormous both during summer and winter; and as but a very small portion of it melts, the accumulated snow is perfectly sufficient to form such a sheet. He does not consider that there is land enough in the south-polar area to supply the astounding number and gigantic size of the icebergs that float in the ocean between lat. 50° and 70°. If this theory be correct, and immense masses of the ice are really afloat, we can easily understand how the whole might, during a southern interglacial period, be broken up, dispersed, and melted by an inflow of equatorial water.

It is quite possible that the ice filling these seas may have originated in pack-ice, which ultimately became converted into a solid and continuous sheet by long ages of successive snowfalls. As layer after layer, converted into ice, was being heaped upon it year by year, the mass would gradually sink until it rested on the sea-bottom.* After this it would assume all the characteristics of continental ice.

* In this opinion I am glad to find that Sir Joseph to a certain extent concurs, for in a letter to me on the subject he says:—"I cannot doubt but that the icebergs have originated from the ice of the great southern barrier; and what I suspect is, that much of this

A theory of the origin of the Antarctic ice, somewhat similar to that of Sir Joseph Hooker, has been advanced by Professor Shaler. In his magnificent work on Glaciers, p. 31, he states his views as follows:—

"When the snow-line touches the sea-level it is because the forces that take away the snow are no longer sufficiently active to overcome the annual accumulation. The existence of such an extremity of cold leads necessarily to the formation, on the surface of the land-locked seas of the circumpolar region, of very thick ice. . . . On the surface of this ice the snows of each winter accumulate and help to increase the thickness of the mass. The ice-floes north of Baffin's Bay, and the straits and inlets that enter the Arctic Sea from the northward, contain a great deal of this ice, which has a thickness of more than 100 feet. In the sea

barrier-ice originated in pack-ice over very shallow bays, increased by successive snowfalls. The quantity of snow that falls in summer is enormous south of latitude 50°–60°. Certainly it fell on half the days of each summer month during the three seasons we spent in those seas, and I think in one month snow fell *every day*. There is no summer melting of snow and ice in the Antarctic as there is in the Arctic regions. It is the only region known to me where there is perpetual snow on land at sea-level."

Now, if the snow which falls in the Antarctic regions at the sea-level does not all melt, but some of it remains year by year, then permanent ice formed at the sea-level, whether it be on frozen pack or on the ground, must be a necessary consequence. If this be so, it cannot be true, as Mr. Wallace affirms, that there is no permanent ice formed but on high land.

Perpetual Snow at the Sea-Level in the Arctic as well as in the Antarctic Regions.—Commander Julius Payer says, "Franz-Josef Land appears even in summer to be buried under perpetual snow, interrupted only where precipitous rock occurs." But, more than this, all the smaller islands are completely covered with separate ice-sheets of their own.—"Austrian Arctic Voyage," vol. ii., pp. 83, 84.

But supposing the snow were not perpetual, this would not prevent, as will be shown in the next chapter, the formation of permanent ice.

into which, or rather on to which, the Nares expedition penetrated, the floe-ice seems to be in a fashion impounded, so that it cannot escape freely to the southern regions. In its prison it appears to continue to drift and grow for ages, so that the name of Paleochrystic Sea, or sea of ancient ice, given it by the officers of the Nares expedition, is well deserved. This mass seems, in fact, to be essentially a floating *névé*, like that which covers Northern Greenland, in everything save the peculiarities that come from its formation on water. Its depth was not accurately determined, but its perfect continuity and vast extent, together with the great irregularities of its surface, make it likely that it exceeds anything in the shape of floe-ice found in the regions known to polar explorers. It seems probable that the so-called Antarctic continent is nothing but an immense sheet of ice, such as this Paleochrystic Sea would become if it were to increase in depth until it fastened on the bottom of the sea. Given a vast sheet of ice, wrapping the surface of a circumpolar sea, supposing it to grow from winter cold and snows more rapidly than the melting of the water could remove it, the result would be that the ice-sheet would in time cleave to the bottom of the sea and become a true glacier, although any portion of its bed was below the level of the water. In view of the southward pointing of the southern continents, and the gradual falling out of land towards the South Pole, this seems to me to be a more likely hypothesis than that which now finds expression in our geographies, where the presence of eternal ice is taken as evidence of a continental development of land in that region. So far, I believe, we have no sufficient evidence of the existence of any other surface than ice above the level of the water in that so-called Antarctic continent."

I have been informed by Capt. Sir Frederick J. O. Evans, Hydrographer of the Admiralty, that, in the compilation of his most instructive and useful Ice Chart of the South Polar regions, he was struck with the remarkable character of the ice foreshores of all the parts of the Antarctic continent sighted by voyagers, and of the undoubted occasional disruption of hundreds or more miles of foreshore ice. The ice, he believes, is evidently formed on comparatively low and level land, and is thrust out in a continuous sheet into deep water, where it breaks up into bergs.

Assuming then, what seems thus probable, that the Antarctic regions consists of low discontinuous land, it will help to explain, as will be shown in a future chapter, the disappearance of the ice during the warm interglacial periods of the southern hemisphere.

On the Argument against the Existence of a South-Polar Ice-cap.—We have certainly no evidence that, during even the severest part of the glacial epoch, an ice-cap, like that advocated by Agassiz and other extreme glacialists, ever existed at the North Pole; I am, however, unable to admit with Mr. Alfred R. Wallace that some such cap, though of smaller dimensions, does not at present exist at the South Pole. Speaking of the Antarctic ice-cap, Mr. Wallace says:—"A similar ice-cap is, however, believed to exist on the Antarctic Pole at the present day. We have, however, shown that the production of any such ice-cap is improbable, if not impossible; because snow and ice can only accumulate where precipitation is greater than melting and evaporation, and this is never the case except in areas exposed to the full influence of the vapour-bearing winds. The outer rim of the ice-sheet would inevitably exhaust the air of so much of its moisture, that what reached the inner parts would produce far less snow than would be melted by the long hot days of summer."*

This opinion, that the mass of ice is probably greatest at the outer rim, which of course is most exposed to moist winds, and that it gradually becomes less and less as we proceed inwards till at last it disappears altogether, is by no means an uncommon one.

It was to establish this conclusion that Professor Nordenskjöld's famous expedition over the ice of Greenland was undertaken.

It by no means follows, as some might be apt to suppose, that the ice must be thickest where the snowfall is greatest. In the case of continental ice, the greatest thickness must always be at the centre of dispersion; but it is here that, owing to distance from the ocean, the snowfall is likely to be least.

We have no reason to believe that the quantity of snow falling, at least at the South Pole, is not considerable. Lieut. Wilkes estimated the snowfall of the Antarctic regions to be about 30 feet per annum; and Sir John Ross says that during a whole month they had only three days free from snow. But there is one circumstance which must tend to make the snowfall near the South Pole considerable, and that is the inflow of moist winds in all directions towards it; and as the area on which these currents deposit their snow becomes less and less as the Pole is reached, this must, to a corresponding extent, increase the quantity of snow falling on a given area. Let us assume, for example, that the clouds in passing from lat. 60° to lat. 80° deposit moisture sufficient to produce, say, 30 feet of snow per annum, and supposing that by the time they reach lat. 80° they are in possession of only

* "Island Life," p. 156.

one-tenth part of their original store of moisture, still, as the area between lat. 80° and the Pole is but one-eighth of that between lat. 60° and 80°, this would, notwithstanding, give 24 feet as the annual amount of snowfall between lat. 80° and the Pole.

However small may be the snowfall, and consequent amount of ice formed annually around the South Pole, unless it all melted it must of necessity accumulate year by year till the sheet becomes thickest there; for the ice could not move out of its position till this were the case. But supposing there were no snow whatever falling at the Pole and no ice being formed there, still this would not alter this state of matters; for in this case, the ice forming at some distance from the Pole, all around would flow back towards the centre, and continue to accumulate there till the resistance to the inward flow became greater than the resistance to the outward; but this state would not be reached till the ice became at least as thick on the poleward as on the outward side. There is no evading of this conclusion unless we assume, what is certainly very improbable, if not impossible, viz., that the ice flowing polewards should melt as rapidly as it advances. We know, however, that in respect to the ice which flows outwards towards the sea, little, if any, of it is melted; and it is only after it breaks off in the form of bergs and floats to warmer latitudes that it disappears, and that even with difficulty. It is therefore not likely that the ice flowing inwards towards the Pole, and without the advantage of escape in the form of bergs, should all happen to melt. If *little or none* of the ice flowing toward the Equator melts, it is physically impossible that *all* the ice flowing polewards should manage to do so; and if it did not all melt, it would accumulate year by year around the Pole till it acquired a thickness sufficient to prevent any further flow in that direction, or, in other words, till its thickness at the Pole became as great as it is all around.

The opinion that the great mass of the ice on the Antarctic continent and also on Greenland lies near to the outer edge, and that it gradually diminishes inwards till at last it disappears, is evidently one based on a misapprehension as to the physical conditions of continental ice. I cannot help believing that had Professor Nordenskjöld duly reflected on the necessary physical and mechanical conditions of the problem, he would not have undertaken the journey across the Greenland ice with the hope of finding green fields in the interior of that continent.

CHAPTER VI.

EXAMINATION OF MR. ALFRED R. WALLACE'S MODIFICATION OF THE PHYSICAL THEORY OF SECULAR CHANGES OF CLIMATE.

Effect of Winter Solstice in Aphelion.—The Star Storage of Cold.—Highland and heavy Snowfall in Relation to the Glacial Epoch.—The only Continental Ice on the Globe probably on Lowlands.—Modification of Theory Examined.—General Statement of the Theory.—Points of Agreement.

THERE are few authors to whom I am more deeply indebted than to Mr. Alfred R. Wallace for his very clear and able exposition and defence from time to time of the main points of the Physical Theory of secular changes of climate. I have read the chapters relating to this subject in his recent work ("Island Life") with a great amount of interest and pleasure; and I need hardly add that, in the main, I agree with the views which he advocates. In these chapters, however, there have been advanced some modifications of the theory which, after a very careful reconsideration of the whole subject, I am wholly unable to accept. As these modifications relate to points of the greatest importance in the question of geological climate, I shall now proceed to examine them at some length. It appears to me, however, that what Mr. Wallace regards as modifications are in some cases really necessary parts of the theory. These may not, it is true, have been in all cases expressed by me, but they are nevertheless implied in the theory. Other points, again, regarded as modifications, are simply facts lying altogether outside of the theory, which can in no way affect it.

Before proceeding, however, to examine in detail Mr. Wallace's modifications of the theory, it may be as well to consider one or two minor points on which I differ from him, as this will save the necessity of referring to them when we come to discuss his main argument.

Effect of Winter Solstice in Aphelion.—At page 126 ("Island Life") he says:—"We may therefore say generally, that during our northern winter, at the time of the glacial epoch, the northern hemisphere was receiving so much less heat from the sun as to lower its surface-temperature on an average about 35° F., while during the height of summer of the same period it would be receiving so much more heat as would suffice to raise its mean temperature about 60° F. above what it is now." In a footnote he adds that "the reason of the increase of summer heat being 60° while the decrease of winter cold is only 35°, is because our summer is now *below* and our winter *above* the average."

There is surely a confusion of ideas here. It is of course true that, as our summer at present occurs in aphelion and our winter in perihelion, the temperature of the former is below and that of the latter above the average; but this can afford no grounds for the result Mr. Wallace attributes to it unless it be assumed (for which there are no astronomical grounds) that our summer is 25° *further* below the average than our winter is above it.

On the Storage of Cold.—In a section on the Effects of Snow on Climate, Mr. Wallace points out the different effects produced by water falling as a liquid in the form of rain and as a solid in the form of snow. The rain, however much of it may fall, runs off rapidly, he states, without producing any permanent effect on temperature. But if snow falls, it lies where it fell, and becomes compacted into a mass which keeps the earth below and the air above at or near the freezing-point. When the snow becomes perpetual, as on the summits of high mountains, permanent cold is the result; and however strong the sun's rays may be, the temperature of both the air and the earth cannot possibly rise much above the freezing-point. "This," he says, "is illustrated by the often-quoted fact that at 80° N. lat. Captain Scoresby had the pitch melted on the one side of his ship by the heat of the sun, while water was freezing on the other side owing to the coldness of the air." Doubtless this is perfectly correct; but on page 502 he states that he has pointed out with more precision than has, he believes, hitherto been done, the different effects on climate of water in the liquid and solid states. This is a somewhat doubtful statement; for in Chapter IV. 'Climate and Time,' in "Phil. Mag." March, 1870, and in other places will, I think, be found all that this section contains. In fact the influence of snow and ice as a *permanent source of cold* is one of the main factors of my theory. The three great factors are (1) the influence of snow and ice, (2) the influence of aqueous vapour, and (3) the influence of ocean-currents. How persistently has it been urged as an objection to my theory that, during the glacial epoch, the great heat of the perihelion summer would more than counterbalance the effect of the aphelion winter. But I have maintained that the summers, notwithstanding the intensity of the sun's rays, instead of being warmer than at present, would in reality be far colder; for this reason, that the temperature of a snow- and ice-covered country can never rise much above the freezing-point. As an example of this I pointed out that, 'were it not for ice, the summers of North Greenland would be as warm as those of England (whereas in point of fact they are colder than our winters); and that were India covered with an ice-sheet, its summers would be colder than those of England.'

"Another point," he says, "of great importance in connection with this subject is the fact that this permanent storing-up of cold depends *entirely* on the annual amount of snowfall in proportion to that of the sun- and air-heat, and not on the actual cold of winter, or even on the average cold of the year." This I have shown at considerable length in Chapter III., is one of the most widespread and fundamental errors within the whole range of geological climatology. Perpetual snow, instead of being due "entirely" to the annual amount of snowfall in proportion to the quantity of heat received by the snow, is in most cases not even *mainly* due to this cause. The overlooking of the fact that in the conservation of snow the temperature of the snow itself is one of the main factors has been a fruitful source of error.

High Land and Heavy Snowfall in relation to the Glacial Epoch.—According to Mr. Wallace, "high land and great moisture" are essential to the initiation of a glacial epoch. Undoubtedly high land and great moisture are the most favourable conditions for bringing about a glacial state of things; but I can hardly agree with him that they are necessary and indispensable.

As to the second of these conditions, great moisture

is required only in order to produce a great snowfall; a great snowfall only in order that the snow may become permanent; and the permanent snow in turn is only in order to have permanent glaciation. But it has already been shown in Chapter III. that we frequently have permanent snow with a very light snowfall, even where the direct heat of the sun is excessive, as on the summits of lofty mountains. Greenland also has but a very small snowfall, and yet the snow and ice are there perpetual. What is necessary is, that the small amount which falls should not all melt. If this be the case, the ice will accumulate year by year, and a glacial condition will ultimately result.

Suppose that the annual precipitation of snow on a continent is equivalent to only 10 inches of ice, and that at the end of each summer one inch remains unmelted, then, in such case, the ice will continue to accumulate year by year until the quantity annually discharged by the outward motion from the centre of dispersion equals that annually formed. But in the case of a continent, this condition can be attained only when the sheet at the centre becomes of enormous thickness. Whether high land be necessary to a glacial epoch or not, it is evident that a heavy snowfall is not an indispensable condition.

As to the first of these conditions, namely High Land, it must be borne in mind that the question is not, Could the causes which are *now* in operation bring about a glacial condition of things without high land? but, Could those physical agencies brought into operation during a high state of eccentricity produce a glacial state of things without high land? Mr. Wallace's answer is that they could not. But I am not satisfied with the grounds on which he bases this opinion. A condition of the greatest importance, though one not absolutely necessary to the production of a glacial epoch, as will presently be shown, is the existence of perpetual snow. The question then is, Could not those physical agencies brought into operation during a high state of eccentricity cover low lands with perpetual snow without the aid of high lands? Mr. Wallace replies, "Perpetual snow nowhere exists on low lands." Supposing this were true (I have endeavoured to show in the last chapter it is not), still it does not follow that perpetual snow may not have existed on low lands, or that, when the present condition of things changes, it may not yet exist. It is not difficult to conceive how, under certain conditions, the snow-line may in some places have been brought to the sea-level. In arctic, or even in subarctic regions, an excessively heavy snowfall, followed by piercingly cold winds from the north, during the whole of the summer months, would keep the snow at a low temperature, and certainly prevent it from disappearing. Keep the surface of the snow at or below the freezing-point, and melting will not take place, no matter how intense the sun's rays may be. A strong wind below the freezing-point will cool the surface of the snow more rapidly than the sun can heat it. Another cause which would tend to keep the snow at a low temperature would be that, along with a cold northerly wind, there is usually a great diminution of aqueous vapour, thus allowing the surface of the snow to radiate its heat more freely into stellar space. For, were it not for the aqueous vapour in the atmosphere, as has been shown in Chapters II. and III., the snow-line, even at the equator, would descend to the sea-level.

Perhaps it is owing to the warm southerly winds of

the two midsummer months that Siberia, even with its inconsiderable snowfall, is not at the present day covered with permanent snow and ice. Mr. Wallace mentions that "in Siberia, within and near the Arctic circle, about six feet of snow covers the country all the winter and spring, and is not sensibly diminished by the powerful sun so long as northerly winds keep the air below the freezing-point, and occasional snow-storms occur. But early in June the wind usually changes to southerly, and under its influence the snow all disappears in a few days." But what would be the consequence were these northerly winds to continue during the whole of June and July? It would probably be that the snow of autumn would begin to fall before that of spring had disappeared. Were this to result, the country would soon become covered with permanent ice. Matters would be still worse if these southerly winds, instead of ceasing, were simply to change from June and July to December and January, for then, in place of producing a melting effect, they would greatly add to the snowfall.

The only Continental Ice on the Globe probably on Lowlands.—The only two continents on the globe covered by permanent ice and snow are Greenland and the Antarctic. But are these continents to be regarded as high lands or as low lands? Mr. Wallace maintains that they are high lands. "It is," he says, "only where there are lofty mountains or elevated plateaus, as in Greenland, &c., that glaciers accompanied by perpetual snow cover the country. The north polar area is free from any accumulation of permanent ice, excepting the high lands of Greenland and Grinnell Land." And in regard to the Antarctic continent he says, "The much greater quantity of ice at the south pole is undoubtedly due to the presence of a large extent of high land." Were it not for these extensive highlands and lofty mountains, Greenland and the Antarctic regions, according to Mr. Wallace's theory, would be free from permanent snow and ice. He, however, nowhere, so far as I can find, offers any proof for the conclusion that those regions possess extensive high lands, elevated plateaus, and lofty mountains sufficient to account for these icy mantles. In the last chapter the subject has been discussed at considerable length, and the conclusions arrived at are diametrically the opposite of those advocated by Mr. Wallace, viz., that Greenland, and probably the greater part of the Antarctic regions, consists of land probably not much above sea-level, and that the mass of ice under which they are buried must be due to some other cause than elevation of the land.

Permanent Ice may originate without Perpetual Snow.—It is not necessary that, in order to have permanent ice, there should be perpetual snow. If snow softens or becomes partially melted and afterwards re-freezes, it is then far more difficult to melt than it was in its original condition. Half-melted snow, when re-frozen, resists the summer sun long after the loose snow has disappeared. We have a good example of this amongst our Scottish mountains. In many places the frozen half-melted snow is permanent; and were the climate from any cause to become deteriorated its amount would yearly increase. In fact, it might go on increasing till not only all our Highlands but the greater part of the surface of Scotland might be covered with ice, long before the snow-line had descended below the level of the summit of Ben Nevis.

Modification of the Theory Examined.

Mr. Wallace's chief, and, I may say, only real modification of my theory is this. I give it in his own words:—

"The alternate phases of precession—causing the winter of each hemisphere to be in *aphelion* and *perihelion* each 10,500 years—would produce a complete change of climate only where a country was *partially* snow-clad; while, whenever a large area became almost *wholly* buried in snow and ice, as was certainly the case with Northern Europe during the glacial epoch, then the glacial conditions would be continued, and perhaps even intensified, when the sun approached nearest to the earth in winter, instead of there being at the time, as Mr. Croll maintains, an almost perpetual spring."—P. 503.

"When geographical conditions and eccentricity combine to produce a severe glacial epoch, the changing phases of precession have very little, if any, effect on the character of the climate, as mild or glacial, though it may modify the seasons; but when the eccentricity becomes moderate and the resulting climate less severe, then the changing phases of precession bring about a considerable alteration and even a partial reversal of the climate."—P. 153.

Again—"It follows that towards the equatorial limits of a glaciated country alternations of climate may occur during a period of high eccentricity, while near the pole, where the whole country is completely ice-clad, no amelioration may take place. Exactly the same thing will occur inversely with mild Arctic climates."—P. 154.

I have, on the contrary, maintained that the more severe the glacial condition of the one hemisphere, the warmer and the more equable would necessarily be that of the other; for the very same combination of causes which would tend to cool the one hemisphere would necessarily tend to warm the other. The process to a large extent consists of a transference of heat from the one hemisphere to the other. Consequently the one hemisphere could not be heated without the other being cooled, nor the one cooled without the other being heated. The hotter the one, the colder the other, and the colder the one, the hotter the other. It therefore follows that the more severe the glacial conditions, the warmer and more equable must be the interglacial warm periods. But, according to Mr. Wallace, there could be no warm interglacial periods, either in temperate or polar regions, except during the commencement and towards the close of a glacial epoch.

Before, however, proceeding to examine in detail the steps by which he arrives at this modification of my theory, it will be as well that the reader should have a clear and distinct knowledge of what that theory really is, and what it professes to explain. These I shall now briefly state in the most general terms, for misapprehension in regard to the main features of the theory lie at the root of most of the objections which have been urged against it.

General Statement of the Theory.—1st. It is not professed that the theory will account for the condition of climate during *all* past geological ages. It treats mainly of the cause of the Glacial Epochs; and one of its essential elements is that these epochs consist of alternate changes, to a greater or less extent, of cold and warm periods; or, in other words, that glacial epochs must consist of alternate glacial and interglacial periods. The chief, though not the sole, aim of the theory is to account for geological climate in so far as such epochs are concerned. Although it could be satisfactorily shown, for example, and this has certainly not yet been done, that during some past geological age, such as the Miocene, the Eocene, or the Cretaceous, the climate was throughout uniformly warm or subtropical, this would not prove that the theory was wrong, unless it could at the same time be shown that the necessary conditions demanded by the theory did then exist. But instead of this supposed condition of climate during Secondary or Tertiary periods being inconsistent with my theory, the fact is, as we shall see by and by, that this theory affords the only rational explanation of such a state of things which has yet been given.

2nd. The theory is not that a high state of eccentricity will necessarily produce a glacial epoch. No misapprehension has been more widespread or more difficult to remove than this. From the very commencement I have maintained that no amount of eccentricity, however great, could produce a glacial condition of things; that the Glacial Epoch was the result, not of a high state of eccentricity, but of a combination of *Physical* Agencies, brought into operation by means of this high state*. As an example of this misapprehension, how frequently has the present condition of the planet Mars been adduced as evidence against the theory. The eccentricity of Mars' orbit is at present greater than that of the Earth's even when at its superior limit; and its southern winter solstice is not far removed from aphelion. It is therefore maintained that, if my theory of the cause of the glacial epoch be correct, the southern hemisphere of Mars ought to be under a glacial condition, and the northern enjoying a perpetual spring—and this, as is well known, is not the case. Here it is assumed that, according to the theory, eccentricity alone ought to

* For this reason I prefer to term the theory the Physical Theory rather than the Eccentricity Theory, as it has been called by some writers.

produce a glacial epoch, irrespective of the necessary physical conditions. We know with certainty that those physical conditions which, according to the theory, were the direct cause of the glacial epoch on our globe, cannot possibly exist on the planet Mars.* Just take one example: either the properties of water on the planet Mars or the conditions of its atmosphere must be totally different from that of our earth; for were our earth removed to Mars' distance from the sun, our seas would soon become solid ice, and we could have neither snow nor rain, ocean-currents, nor any of the necessary conditions for secular change of climate. This is doubtless not the present state of Mars; but the reason of this can only be that the physical and meteorological conditions of the planet must be wholly different from those of the earth.

When we reflect that a very slight change in the properties of aqueous vapour, or in the condition of our atmosphere, would effectually prevent the possibility of a glacial epoch occurring on our earth, notwithstanding a high state of eccentricity, we need not wonder that the planet Mars is not in a state of glaciation. But the eccentricity of Mars, though high, is still far from its superior limit, and the planet may yet, for any thing which we know to the contrary, pass through a glacial epoch.

3rd. Another prevailing misapprehension is the supposition that the theory does not recognize the necessity for geographical conditions. In reading "Island Life" one might be apt to suppose that one of the chief points of difference between Mr. Wallace and myself is that he regards geographical distribution of sea and land as an important factor in a theory of geological climate, whereas I entirely ignore this

* See 'Climate and Time,' p. 79.

condition. Nothing could be further from the truth than such a supposition. I can boldly affirm that the necessity for geographical conditions is as truly a part of my theory as of Mr. Wallace's modification thereof.

One of the most important agencies, according to my view, is the enormous amount of heat conveyed from equatorial to temperate and polar regions by means of ocean-currents, and the deflection of this heat, during a high state of eccentricity, from the one hemisphere to the other. But all this depends on ocean-currents flowing from equatorial to polar regions; and the existence of these currents in turn depends, to a large extent, on the contour of the continents and the particular distribution of sea and land. Take, as one example, the Gulf-stream, a current which played so important a part in the phenomena of the glacial epoch. A very slight change in geographical conditions, such as the opening of communication between the Gulf of Mexico and the Pacific, would have greatly diminished, if not entirely destroyed, that stream. Or, as I showed on a former occasion, a change in the form or contour of the north-east corner of the South-American continent would have deflected the great equatorial current, the feeder of the Gulf-stream, into the Southern Ocean and away from the Carribean Sea. One of the main causes of the extreme condition of things in North-western Europe, as well as in eastern parts of America, during the glacial epoch, was a large withdrawal of the warm waters of the Gulf-stream; and this was to a great extent due, as I stated in my very first paper on the subject *, to the position of Cape St. Roque, which deflected the equatorial current into the Southern Ocean. That a

* "Phil. Mag," for August, 1864.

geographical distribution of land and water permitting of the existence and deflection of those heat-bearing currents is one of the main factors in my theory is what must be obvious to every reader of 'Climate and Time.'

The difference between Mr. Wallace and myself is this:—I maintain that with the *present* distribution of land and water, without calling in the aid of any other geographical conditions than now obtain, those physical agencies detailed in 'Climate and Time' are perfectly sufficient to account for all the phenomena of the glacial epoch, including those intercalated warm periods, during which Greenland would probably be free from ice and the Arctic regions enjoying a mild climate; while Mr. Wallace, on the other hand, maintains that without assuming some *change* in the geographical conditions of our globe those physical agencies will not account for that state of things, at least in so far as the disappearance of the ice in Arctic regions is concerned.

To narrow the field of inquiry, and bring more prominently before the mind the real question at issue, I shall state the main points on which Mr. Wallace and I appear to agree.

Points of agreement.—1. Mr. Wallace agrees with me that a high state of eccentricity could never directly produce a glacial condition of climate; that the glacial epoch was the direct result, not of a high state of eccentricity, but of a combination of physical agencies brought into operation by means of this high state.

2. He agrees with me also in regard to what these physical agencies really were; for the agencies to which he refers in his "Island Life" are almost identically those which I have advanced in 'Climate and Time' and elsewhere.

3. Mr. Wallace agrees with me in regard to the mutual reactions of the physical agents. He maintains with me that these physical agencies not only all lead to one result—the accumulation of snow and ice—but that their efficiency in bringing about this result is strengthened by their mutual reactions on one another. At pp. 137-139 he gives a variety of examples of these mutual reactions, and says that they "produce a maximum of effect which, without their aid, would be altogether unattainable."

4. As has already been shown, we both agree as to the necessity of certain geographical conditions for the production of the glacial epoch. For although that epoch was mainly brought about by the physical agencies, yet these agencies could not have produced the required effect unless the necessary geographical conditions had been supplied, these being necessary for their effective operation.

5. Mr. Wallace admits, of course, that the necessary geographical conditions existed during the glacial epoch; for, unless this had been the case, no glacial epoch could have occurred. Therefore, all that was required to produce glaciation was an amount of eccentricity sufficient to set the physical agencies into operation. Be it observed, it did not require, in *addition* to the physical agencies, some changes in the geographical conditions, or some new conditions; for the geographical conditions being existent, all that was then required to bring about the glacial epoch was the operation of the physical agencies. The overlooking of this fact has led to much confusion. For example, 210,000 years ago, with winter in aphelion, "the problem to be solved," says Mr. Wallace, "is, whether the snow that fell in winter would accumulate to such an extent that it would not be melted in summer, and so

go on increasing year by year till it covered the whole of Scotland, Ireland, and Wales, and much of England. Dr. Croll and Dr. Geikie answer without hesitation that it would. Sir Charles Lyell maintained that it would only do so when geographical conditions were favourable." * Here we have a complete misapprehension of the relation between Sir Charles Lyell's views and mine; for I would certainly maintain (and, I presume, Dr. Geikie also) as emphatically as Sir Charles could do, "that it would only do so when geographical conditions were favourable." For undoubtedly, according to the theory advocated in 'Climate and Time,' no glacial epoch could result without geographical conditions suitable for the operation of physical agencies; and this is virtually what Sir Charles maintains. The glacial epoch resulted during the last period of high eccentricity because the geographical conditions suitable for the effective operation of the physical causes then existed.

6. It is assumed in Climate and Time' that, with the exception of those resulting from oscillations of sea-level, afterwards to be considered, the general distribution of sea and land, and other geographical conditions, were the same during the glacial epoch as they are at present.† Consequently, in accounting for the glacial epoch I had only to consider the effects

* "Island Life," p. 136.

† Prof. J. Geikie, however, believes that during early Postglacial times a considerable change in the physical geography of the North seas took place (see "Prehistoric Europe," chap. xxi.). In order to account for the floras of Greenland, Iceland, and the Faröe Islands, he thinks a land connection must have existed between these places and Scandinavia. For reasons which will be stated in a future chapter I am somewhat doubtful on this point. There is, I think, an important agent overlooked in the question of the distribution of Arctic flora and fauna. Prof. Geikie, however, does not believe that the climatic condition of that period was in any way due to this change.

resulting from those physical agencies called into operation by an increase of eccentricity. To have speculated on hypothetical geographical conditions different from those which now obtain, and on the influence which these may have had in bringing about the glacial epoch, would have been on my part perfectly absurd, as I knew we had no evidence of the existence of any such conditions. Besides, my aim was to account for that epoch from known and established facts and principles, without the introduction of hypothetical causes. I fear that the fact of my making little or no allusion to geographical conditions in my explanations may have unfortunately led Mr. Wallace and others to conclude that I altogether ignore, or, at least, undervalue their importance, which is certainly not the case.

Although Mr. Wallace so frequently alludes to the importance of geographical conditions, I am not sure if he believes that during the glacial epoch those conditions differed materially from what they are at present, or that glaciation could have been greatly influenced by any difference which did exist.

7. Mr. Wallace alludes to one or two geographical conditions which, *if* they had existed during the glacial epoch, would have greatly aided glaciation; as, for example, if a land-barrier had extended from the British Isles, across the Faröe Islands and Iceland to Greenland, cutting off from Northern Europe the warm waters of the Atlantic, including the Gulf Stream. "The result," he says, "would almost certainly be that snow would accumulate on the high mountains of Scandinavia till they became glaciated to as great an extent as Greenland."

It would be easy to multiply cases of this kind where a distribution of land and water different from the present might have been more favourable to glaciation than the present; but the question is, Did any such difference favouring glaciation *actually exist* during the glacial epoch? I have never been able to find any evidence that it did. Many a change in geographical conditions has taken place during Tertiary times, some of which were doubtless favourable to glaciation; but have we any evidence that during the glacial epoch the geographical conditions were more favourable than they are at present? Unless this can be shown to be the case, there is no necessity for referring to a difference in geographical conditions during that epoch as a cause of glaciation. This being so, it does not follow, because in my explanation of the cause of the glacial epoch I may not, like Sir Charles Lyell and others, have speculated on the effects which might have resulted had the distribution of land and water been different from what it is now, that I ought on this account to be charged with undervaluing the importance of geographical conditions.

Trusting that these preliminary considerations may tend to remove the partial confusion in which this somewhat complex subject has been involved, I shall now proceed to examine Mr. Wallace's main argument. I shall consider it, first, in relation to physical principles, and, secondly, in relation to geological and palæontological facts.

CHAPTER VII.

EXAMINATION OF MR. ALFRED R. WALLACE'S MODIFICATION OF THE PHYSICAL THEORY OF SECULAR CHANGES OF CLIMATE.—*Continued.*

Physics in relation to Mr. Wallace's Modification of the Theory.—Another Impossible Condition Assumed. — A Geographical Change not Necessary in Order to Remove the Antarctic Ice.—Other Causes than Antarctic Ice affecting the Northward-flowing Currents. — Climatic Conditions of the Two Hemispheres the reverse Ten Thousand Years ago; argument from.—Mutual Reactions of the Physical Agents in Relation to the Melting of the Ice.—Another Reason why the Ice does not Melt.

I. *Physics in Relation to Mr. Wallace's Modification of the Theory.*

The grand modification, that during the height of the glacial epoch the snow and ice would not disappear when precession brought the winter solstice round to perihelion, I have already given in Mr. Wallace's own words. As the reasons which he assigns for this modification are very briefly stated by him, I may here give them also in his words.

After describing the state of North-Eastern America and the North Atlantic, to which I have already alluded, he says:—

"But when such was the state of the North Atlantic (and, however caused, such *must* have been its state during the height of the glacial epoch), can we suppose that the mere change from the distant sun in winter and near sun in summer, to the reverse, could bring about any important alteration—*the physical and geographical causes of glaciation remaining unchanged?* For, certainly, the less powerful sun of summer, even though lasting somewhat longer, could not do more than the much more powerful sun did during the phase of summer in *perihelion*, while during the less severe winters the sun would have far less power than when it was equally near and at a very much greater altitude in summer. It seems to me, therefore, quite certain that whenever *extreme* glaciation has been brought about by high eccentricity combined with favourable geographical and physical causes (and without this combination it is doubtful whether *extreme* glaciation would ever occur), then the ice sheet will *not* be removed during the alternate phases of precession, so long as these geographical and physical causes remain unaltered. It is true that the warm and cold oceanic currents, which are the most important agents in increasing or diminishing glaciation, depend for their strength and efficiency upon the comparative extents of the northern and southern ice-sheets, but these ice-sheets cannot, I believe, increase or diminish to any important extent unless some geographical or physical change first occurs." *

Again,—"It is quite evident that during the height of the glacial epoch there was a combination of causes at work which led to a large portion of North-Western Europe and Eastern America being buried in ice to a greater extent even than Greenland. Among these causes we must reckon a diminution of the force of the Gulf Stream, or its being diverted from the north-western coasts of Europe; and what we have to consider is, whether the alteration from a long cold winter and short hot summer, to a short mild winter and long cool summer would greatly affect the amount of ice *if the ocean-currents remained the same.* The force of these currents is, it is true, by our hypothesis modified by the increase or diminution of the ice in the two hemispheres alternately, and they then react upon climate; but they

* "Island Life," p. 150.

cannot be thus changed till after the ice accumulation has been considerably affected by other causes."*

There are some further reasons assigned, which will be considered as we proceed.

From what has already been shown, it will be seen that the causes which led to the glacial epoch may be classed under three distinct groups:—(1) the astronomical, (2) the physical, and (3) the geographical. This threefold division is distinctly recognised by Mr. Wallace in the above quotations, as well as in all his reasoning on the subject of geological climate.

In the astronomical group the main elements are the two following:—1st, A high state of eccentricity producing, on the hemisphere whose winter solstice happens to be in aphelion, a long and cold winter with a short and hot summer, and on the other hemisphere, whose winter solstice, of course, at the time is in perihelion, a short and mild winter with a long and cool summer; 2nd, Precession, transferring these conditions from the one hemisphere to the other alternately every 10,000 or 12,000 years. The physical elements are, of course, the influence of snow and ice, ocean-currents, aqueous vapour, clouds, fogs, and a host of other things which have already been discussed at length in Chapters II. and III.; while the geographical consist of the particular distribution of land and water, elevations or depressions in the sea-bottom, contour of the sea coast, and other geographical conditions influencing the flow of ocean-currents.

It is to the influence of physical agencies, however, that the glacial epoch is more directly due. The main function of the astronomical agents is to set and keep the physical agencies in operation, and also to *deter-*

* "Island Life," p. 148.

mine the character of their operations. For example, the position of the winter solstice in relation to the aphelion or to the perihelion, during a high state of eccentricity, determines whether the physical agencies will produce on a given hemisphere a glacial or a warm condition of climate; while precession determines which of the two hemispheres shall be the glaciated and which the warm. In one respect we may say that the astronomical causes produce glaciation by means of the physical agencies.

The geographical conditions, however, cannot properly be considered to be causes in the sense in which the astronomical and physical are. They are more properly *conditions* to the production of a glacial epoch than *causes*. They cannot be said to *act* in the production of glaciation. They are rather permanent and passive conditions enabling the active causes to produce their required effects. Had the glacial epoch resulted from the elevation of the land, as some geologists suppose, then this *elevation* might properly be said to have been the cause of the glacial epoch; but the glacial epoch was produced by no such means, nor by any *change* in the physical geography of the globe. A certain geographical condition of things was, of course, requisite in order to the effective operation of the astronomical and physical causes. This condition existed at the time of the glacial epoch; and it is only in this sense that that epoch can be referred to anything geographical.

It is true that a cause, as Sir William Hamilton states, may be defined as "all that without which the effect would not happen;" but this is far too general an expression of cause for practical purposes. We therefore fix on the particular antecedent or antecedents, through the activity of which the event is mainly brought about, and term them the *causes* of the event, and the others the *necessary conditions.*

I cannot help thinking that the way in which geographical conditions are spoken of as causes of the glacial epoch has tended to confusion.

During the glacial epoch there were frequent submergences and elevations of the land, or rather oscillations of sea-level, and these, it is true, would produce a change in the relative extent of sea and land. But whether we suppose it to have been the sea which rose and fell in relation to the land, or the land in relation to the sea, it equally follows that the geographical change resulting therefrom could not possibly have been a cause of the glacial epoch. It is now a well-established fact that submergence accompanied glaciation; the glaciation may have been that which led to the submergence; but it could not possibly have been the submergence which led to the glaciation. An elevation of the land would have favoured glaciation, but submergence would not. Its tendency would rather be in the opposite direction. It is now also established, that during the continental period, or period of elevation, the climate was warm and equable; for it was then, as has been remarked, that this country was invaded by tropical and subtropical animals. Now, it is equally plain that the elevation could not have been the cause of the heat. Elevation of the land might produce cold, but it could not have been a cause of the heat. It follows, therefore, that the geographical change resulting from submergence or elevation of the land cannot be regarded as a cause of the glacial epoch; for its effect on climate, if it had any, was in opposition to that of the astronomical and physical agencies. It would prove a hindrance, not a help.

Referring now to Mr. Wallace's argument: When glacial conditions in the North Atlantic attained their maximum development, "can we suppose," he asks, "that the mere change from the distant sun in winter, and near sun in summer to the reverse, could bring about any important alteration—the physical and geographical causes of glaciation remaining unchanged?" Here, to begin with, we have an impossible state of things assumed. It is assumed in this question that it is possible for the winter solstice to pass from aphelion to perihelion, and the *physical* causes to remain unchanged. It is assumed as possible that the astronomical conditions might be reversed without a reversal of the physical conditions.

When the winter solstice is in aphelion, it sets in operation many physical causes, the tendency of which is to produce an accumulation of snow and ice; but when the solstice-point moves round to the perihelion, the tendency of these causes is reversed, and they then undo what they had previously done—melt the snow and ice which they had just produced. Now, what Mr. Wallace asks is this: When, owing to the winter solstice being in aphelion during a high state of eccentricity, a glacial condition of things is produced, will the fact of the solstice-point being moved round to perihelion remove the glacial condition, if *the physical causes remain unchanged in their mode of operation?* My reply is, it certainly would not. Here it is assumed that the physical causes are working in opposition to the astronomical; that when the solstice is in perihelion the action of the physical causes, instead of being reversed, as it should be according to theory, still continues to produce and maintain a glacial state of things, the same as it did when the solstice-point was in aphelion; and he asks, Will the astronomical causes in this struggle manage to overpower the physical and produce a melting of the ice? I unhesitatingly reply, No; for the physical causes are far more powerful than the astronomical. The astronomical causes, as we have seen, are perfectly unable to produce a glacial state of things without the *aid* of the physical. How, then, could we expect that they could remove this glacial state if the physical causes were actually working against them?

In thus setting the physical causes against the astronomical, Mr. Wallace is basing his argument for the non-disappearance of the snow and ice on a state of things which cannot possibly, under the circumstances, exist. His question, to have consistency, should be this:—When glacial conditions were at their height, &c., can we suppose that the mere change from the distant sun in winter and the near sun in summer, to the reverse, could bring about any important alteration —the *geographical* causes of glaciation remaining unchanged? If the question is put thus, and it is the only form in which it can be put to be consistent with the theory which Mr. Wallace himself advocates, then my reply is, That the change from the distant sun in winter and near sun in summer to the near sun in winter and distant sun in summer, aided by the change in the physical causes which this would necessarily bring about, would certainly be sufficient to cause the snow and ice to disappear without any change in the geographical condition of things. The combined influence of the astronomical and physical causes, when the winter solstice is in perihelion, is perfectly sufficient to undo all that they had previously done when the solstice was in aphelion. When the action of the causes is reversed, the effects will be reversed.

Had the glacial epoch been produced by geographical

causes, then it is probable that the ice would not have disappeared till these causes were changed. Had the ice, for example, been simply due to an elevation of the land, as some have argued, then it would not probably have disappeared till the land became lowered. But it was the result of no such cause. It was due, not to an elevation of the land, but to a number of physical causes, brought into operation by a high state of eccentricity. This Mr. Wallace fully admits and maintains. A certain geographical state of things was, of course, necessary to enable the astronomical and physical causes to produce the required effect; and this was really all that geographical conditions had to do in the matter. Let this be observed, however, that the *same geographical condition of things* which favours the accumulation of ice when the winter solstice is in aphelion, favours its disappearance when the solstice is in perihelion. This is obvious, because the same combination of physical agencies which makes the hemisphere in aphelion cold, makes the one in perihelion warm. The heating of the one is, to a large extent, the result of the cooling of the other. It is the *transference* of heat by ocean-currents from the hemisphere in aphelion to the one in perihelion which is a main reason why the former is cold and the latter warm. Hence a change in geographical conditions is unnecessary for the disappearance of the ice on the hemisphere with the perihelion winter, whether that hemisphere be the northern or the southern.

The tendency of the combined influence of all the causes—astronomical, physical, and geographical—is to cool the one hemisphere and to warm the other, to accumulate the ice on the one, and remove it from the other. Consequently the same total combination of causes which will produce an accumulation of ice on either hemisphere when the winter solstice is in aphelion will produce a melting of that ice when the solstice moves round to the perihelion.

Another Impossible Condition Assumed.—"What we have to consider," says Mr. Wallace, "is whether the alteration from a long cold winter and short hot summer, to a short mild winter and long cool summer, would greatly affect the amount of ice *if the ocean-currents remained the same.*" Here, again, we have an impossible state of things assumed. It is assumed that, notwithstanding the change from an aphelion to to a perihelion winter, the ocean-currents would still remain the same. And it is asked, would the astronomical causes in this case remove the glaciation? I would be disposed to say that they would not.

"The force of these currents," he adds, "are, it is true, by our hypothesis modified by the increase or diminution of the ice in the two hemispheres alternately (they depend for their strength and efficiency upon the comparative extent of the northern and southern ice-sheets), and they then react upon climate; but they cannot be thus changed till after the ice-accumulation has been considerably affected by other causes."

What, then, are the other causes which affect the ice-accumulation and thus lead to a change in the ocean-currents? "These ice-sheets cannot, I believe," says Mr. Wallace, "increase or diminish to any important extent unless some *geographical* or *physical* change first occurs." The first thing required to affect the ice-accumulation is thus a geographical or a physical change. But we have just seen that the character of the physical causes depends upon the astronomical. A change from a long cold winter and short hot summer to a short mild winter and long cool summer would reverse the operations of the physical causes and lead to a melting of the ice. The physical causes, therefore, offer no barrier. What more do we still require? This we have in the following foot-note:—"The ocean-currents are mainly due to the difference of temperature of the polar and equatorial areas combined with the peculiar form and position of the continents, and some one or more of these factors must be altered *before* the ocean-currents towards the North Pole can be increased." *

One of these factors—change in the form and position of the continents—may be left out of consideration; for we have no evidence of any such change during the glacial epoch, except one, which, as has been already proved, could have had no effect. We must, therefore, look to a change in "the difference of temperature of the polar and equatorial areas" for any increase in the currents towards the North Pole. And in order to bring about this change, "the only available factor," Mr. Wallace states, "is the Antarctic ice; if this were largely increased, the northward-flowing currents might be so increased as to melt some of the Arctic ice. But without some geographical change the Antarctic ice could not materially diminish during its winter *perihelion,* nor increase to any important extent during the opposite phase. We therefore seem to have no available agency by which to get rid of the ice over a glaciated country, so long as the *geographical conditions* remained unchanged and the eccentricity continued high."

According to Mr. Wallace, the *only* available factor to produce a difference of temperature between the

* "Island Life," p. 150.

south-polar area and the equator, so as to increase the north-flowing currents and thus melt the Arctic ice, would be an increase of the Antarctic ice; but this he considers impossible without some *geographical change.* Without such a change, the Antarctic ice, he maintains, would neither be increased nor diminished. Hence it follows that without this change there is, according to Mr. Wallace's theory, no possibility of getting quit of our northern ice during interglacial periods.

This sweeping conclusion seems to be based on two assumptions, both of which appear to me to be erroneous. *First,* that the "only" factor available is the Antarctic ice; and, *secondly,* that the Antarctic ice can neither be increased nor diminished without some geographical change.

A Geographical Change not Necessary in order to Remove the Antarctic Ice.—In reference to the first—that the Antarctic ice is the "only" available factor—I shall presently show that there are other causes affecting the northward-flowing currents as powerfully as the Antarctic ice. As to the second—that the Antarctic ice can neither be increased nor diminished materially without some geographical change—this is an assumption based, no doubt, on the opinion which he holds that the Antarctic ice is due to the elevated nature of that continent. Of course, if this opinion be correct, then, without a lowering of the land, the ice can never disappear or be greatly changed in amount by astronomical or physical causes. But from what has already been stated in Chapter V. in reference to the condition of the Antarctic regions, I think it likely that they probably consist of low dismembered land or of groups of flat islands little elevated above sea-level, but all fused together by one continuous sheet of ice. In fact, it seems highly probable that a very large portion of the ice rests on a surface which is under the sea-level. Victoria Land is, of course, certainly elevated and mountainous, but the character of the Antarctic icebergs shows that this state of things must be the exception and not the rule in those regions.

If this be the case, the Antarctic ice is just in the condition admitting of its being easily modified by warm currents from equatorial regions. In fact, at the very present day, as Dr. Neumayer has shown, the slight southward deflections of the warm westerly drift-current caused by the projecting land masses of Australia, Africa, and South America cut notches in the ice. When the southern winter solstice was in perihelion during the glacial epoch, it is probable that the greater part of the ice then disappeared.

In fact, this is a result which would be even still more likely to occur were the views held by Sir Joseph Dalton Hooker, Professor Shaler, and others, detailed in Chapter V., as to the nature of the Antarctic ice, proved to be correct, viz., that the greater part of that ice originated in pack or sea ice, which ultimately became converted into a solid and continuous sheet by long ages of successive snowfalls.

If such be the condition of the Antarctic ice, we can readily understand how it might all soon disappear under the influence which would be brought to bear upon it were the eccentricity high and the southern winter solstice in perihelion. The warm and equable conditions of climate which would then prevail, and the enormous quantity of intertropical water carried into the Southern Ocean, would soon produce a melting of the ice. Layer after layer would disappear off the surface, and as soon as the weight of the sheet became less than that of the water which it had displaced, the sheet would float. After this it would no doubt shortly break up and become dispersed.

Other Causes than Antarctic Ice affecting the Northward-flowing Currents.—If we consider the effect which the present amount of eccentricity, small as it is, has on the climatic condition of some parts of the southern hemisphere, we shall readily understand how, during the glacial epoch, the warm water of this hemisphere may have been impelled northward, even independently of the influence of the Antarctic ice. In order to show the present effect of eccentricity on climate, I cannot do better than quote Mr. Wallace's own words on the subject. Referring to its effect on south temperate America, he says:—

"Those persons who still doubt the effect of winter in *aphelion* with a high degree of eccentricity in producing glaciation, should consider how the condition of south temperate America at the present day is explicable if they reject this agency. The line of perpetual snow in the southern Andes is so low as 6000 feet in the same latitude as the Pyrenees; in the latitude of the Swiss Alps, mountains only 6200 feet high produce immense glaciers which descend to the sea-level; while in the latitude of Cumberland, mountains only from 3000 to 4000 feet high have every valley filled with streams of ice descending to the sea-coast and giving off abundance of huge icebergs. Here we have exactly the condition of things to which England and Western Europe were subjected during the latter portion of the glacial epoch, when every valley in Wales, Cumberland, and Scotland had its glacier; and to what can this state of things be imputed, if not to the fact that there is now a moderate amount of eccentricity, and the winter of the southern hemisphere is in *aphelion?* The mere geographical position of the southern extremity of America does not seem especially favourable to the production of such a state of glaciation. The land narrows from the tropics southwards, and terminates altogether in about the latitude of Edinburgh; the mountains are of moderate height; while during summer the sun is three millions of miles nearer, and the heat received from it is equivalent to a rise of 20° F. as compared with the same season in the northern hemisphere." *

In a similar glacial condition are the islands of South Georgia, South Shetland, Graham Land, Enderby Land, Sandwich Land. There can be little doubt that the present extension of ice in the Antarctic regions is to a considerable extent due also to the influence of eccentricity.

Let us now glance for a moment at the influence which this state of things has at present on northward-flowing currents. One result is that the south-east trades are stronger than the north-east, and as a consequence blow over on the northern hemisphere ten or fifteen degrees beyond the equator. This has the effect, as has been shown ('Climate and Time,' Chapters V. and XIII., and other places), of impelling the warm surface-water of the southern intertropical regions over on the northern hemisphere. It is possible that the greater strength of the south-east trades may to some extent be due to the preponderance of ocean on the southern hemisphere; but there can be little doubt that it is mainly the effect of eccentricity.

The result of this transference of water from the southern to the northern hemisphere is that the intertropical waters of the northern hemisphere are between three and four degrees warmer than those of the

* "Island Life," p. 142.

southern. Another result which follows, as has also been shown, is that the great equatorial currents are made to lie at some distance to the north of the equator; hence, when they are impelled against the American and the Asiatic continents, and become deflected northwards and southwards, the larger portion of the water goes to the north, and thus raises the temperature of the northern hemisphere. Now, if all this results as a consequence from the present small amount of eccentricity, how much greater must have been the effect during the glacial epoch, when the eccentricity was more than three times its present value, and the southern winter also, as now, in aphelion! All those effects which we have just been considering would then have been magnified far more than threefold.

Climatic Conditions of the Two Hemispheres the Reverse 10,000 or 12,000 years ago: Argument from.—Ten or twelve thousand years ago, when our northern winter solstice was last in aphelion, the climatic conditions were in all probability the reverse of what they are at present. There appears to be pretty good geological evidence that such was the case. This, under the present small amount of eccentricity, shows not only to what an extent climate is affected by eccentricity, but also (and with this we are at present more particularly concerned) that its tendency is to cool the one hemisphere and warm the other, to accumulate the snow and ice on the one and melt them on the other. And this result, to a large extent, is doubtless brought about by its influence on ocean-currents.

There are good reasons for concluding, as Professor J. Geikie has fully shown,* that at a very recent date

* "Prehistoric Europe," p. 411.

(during the time of the formation of the 40-feet raised beach and the deposition of the Carse-clays) the climate was much colder than it is at present. The seas surrounding our island appear to have had a lower temperature than they have at present; and our Highland valleys seem to have been occupied by local glaciers.*

The Carse-clays of Scotland are best developed in the valleys of the Tay, the Earn, and the Forth. These deposits consist of finely laminated clays and silt. "Now and again," says Professor J. Geikie, "the deposits consist of tough tenacious brick-clay, which does not differ in appearance from similar brick-clays of glacial age." The clay is usually free from stones, but occasionally blocks of six inches or a foot in diameter are found in it; and Professor J. Geikie mentions having seen one four feet in thickness. Stones of this size in a fine laminated clay evidently indicate the presence of floating ice. But, as Professor J. Geikie remarks, "it is rather the general character of the clays themselves than the presence of erratics which indicates colder climatic conditions. The fine tenacious brick-clays are not like the dark sludge and silt which now gather upon the estuarine bed of the Tay, but resemble and in some cases are identical in character with the laminated clays of true glacial age with Arctic shells." These Carse-clays, as he further remarks, appear in a large measure to be made up of the fine "flour of rock" derived from the grinding action of glaciers

* In a paper "On the Obliquity of the Ecliptic," read before the Geological Society of Glasgow in 1867, I concluded that at the time of the deposition of the Carse-clays the mean winter temperature was probably 10° or 15° lower than at present, and the Gulf Stream considerably reduced. See also 'Climate and Time,' pp. 403-410.

which then occupied the Highland valleys, and from which muddy waters escaped in large quantities in summer, owing to the melting of the snow and ice. In short, these Carse-clays appear to coincide with the most recent period of local glaciers.

During that period some of the glaciers, as Professor J. Geikie has shown, appear to have even reached the sea-level. For example, at the mouth of Glen Brora, in Sutherland, there is a well-marked moraine with large blocks resting upon, and apparently of the same age as, the deposits of the raised beach.* Mr. Robert Chambers also observed moraine matter resting upon the 30-feet beach at the opening of Glen Iorsa, in Arran. In many of the Highland sea-lochs, says Professor J. Geikie, glaciers appear to have come down to the sea and calved their icebergs there. This, he thinks, is probably the reason why the 40–50–feet beach is not often well seen at the heads of such sea-lochs. The glaciers seem in many cases to have flowed on for some distance into the sea, and thus prevented the formation of a beach and cliff-line.

The greater magnitude and torrential character of the rivers of that period were no doubt due to the melting during summer of great masses of snow and ice. The presence of the large Greenland whale, found frequently in the Carse deposits, would seem to indicate a somewhat colder sea than now surrounds our island. A decrease of temperature of the sea is what would necessarily occur from a slight diminution in the volume of the Gulf Stream, arising from the greater deflection of equatorial water into the southern hemisphere.

Another circumstance deserves notice here, as it

* "Prehistoric Europe," p. 411.

seems to indicate that the climatic conditions of the two hemispheres were at the period of the Carse-clays the reverse of what they are at present. During that period the sea stood higher in relation to the land than it does at the present time. To this circumstance alone no great importance can be attached; but when we consider, in addition, that submergence has almost invariably accompanied glaciation, we may regard it as highly probable that the submergence at the period in question was the result of a greater amount of ice on the northern hemisphere and a less amount on the southern than now. This probability is further increased by the fact that during the growth of the ancient Forest, which immediately underlies the Carse-clays, and indicates a condition of climate even more warm and equable than the present,* the sea stood not only higher in relation to the land than it did during the time of the deposition of the Carse-clays, but somewhat higher than it does at present. The buried Forest, doubtless, belongs to the period 10,000 or 12,000 years prior to that of the Carse-clays,† when the winter solstice was in perihelion; and at this time, owing to a somewhat greater amount of eccentricity than at present, the quantity of ice on the southern hemisphere might be expected to be greater, and that on the northern less, than now.

Thus, when the northern winters were last in aphelion there was a *rise* of sea-level, resulting, doubtless, from a preponderance of ice on the northern hemisphere; but when the buried Forest flourished, 10,000 or 12,000 years prior, the winters

* Those who doubt the equable and warmer character of the climate of the submarine Forest-bed period should study the mass of evidence on this point given in "Prehistoric Europe."

† For the probable dates of the Carse-clays and the submarine Forest-beds, see 'Climate and Time,' p. 407.

were in perihelion, and there was a *fall* of sea-level, due, in all likelihood, to the preponderance of ice on the southern hemisphere. But this is not all: the strata which underlie the buried Forest bear witness to another *rise* of sea-level.

These changes of climatic conditions and oscillations of sea-level, which took place during the latter part of the Postglacial period, are just what should have taken place on the supposition that they were the result of those astronomical and physical agents which we have been considering. Thus, immediately preceding the Present period, we have that of the 25- and 40-feet* raised beaches and the Carse deposits, which indicate that the climate was then more severe and the sea somewhat colder and standing at a higher level than at present. Now, during this Recent period, our northern winter solstice was in aphelion, and the condition of things is exactly what, according to theory, we ought to expect.

Preceding the period of the Carse-clays came that of the buried Forest, when the climate was even more genial and equable than at the present day, the Gulf Stream larger and the sea at a lower level than now. Now, during this period, the winter solstice was in perihelion, and the eccentricity somewhat greater than at present; and here again we have exactly that condition of things which, according to theory, we ought to expect. It would be very singular indeed were there no physical connection between these conditions and the causes to which I have been attributing them. It would certainly be singular were all these coin-

* At one time I thought ('Climate and Time,' p. 409) that the 40-feet beach might belong to a period 50,000 years prior to the Carse-clays; but I am now satisfied that the two beaches both belong to the period of the Carse-clays, as Professor J. Geikie has shown in "Prehistoric Europe," Chap. XVI.

cidences purely accidental. These changes have all been so recent, geologically speaking, and so general and widespread in their character, that they cannot reasonably be attributed to any known geographical changes. If we admit, then, that they were the result of those astronomical and physical agents to which I have referred them, we must also admit that those agents were as efficient in producing a warm and equable climate as in producing a cold and severe one. We must further admit that, with a very small amount of eccentricity, widely marked differences of climatic conditions are brought about on the two hemispheres; that, when the winters are in perihelion, the melting of the snow and ice and the increase of the Gulf Stream and other northward-flowing currents are as necessary a result as were the formation of the snow and ice and the decrease of the Gulf Stream and those currents when the winters were in aphelion. And if this holds true in reference to recent and postglacial times, when the eccentricity was small, it must, for reasons which will presently be stated, hold true in a higher degree in reference to the glacial epoch, when the eccentricity was more than three times its present value.

The Mutual Reaction of the Physical Agents in Relation to the Melting of the Ice.—When the winter solstice is in aphelion it sets in operation, according to theory, as has been shown, a host of physical causes the tendency of which is to produce an accumulation of snow and ice; but when the solstice-point moves round to perihelion the tendency of these causes is reversed, and they then undo what they had previously done—they melt the snow and ice which they had just produced. The action of the causes being reversed, the effects are reversed. But it must be observed that the greater the amount of the eccentricity, the greater will be the effect resulting from the combination of these physical agents, whether that effect be the production of snow and ice on the cold hemisphere, or the melting of them on the warm,—whether it be their production when the winter solstice of a hemisphere is in aphelion, or their melting when that solstice is in perihelion.

We have, however, to take into account not merely the action of the physical agents, but their mutual reactions on each other. The effect of this mutual reaction is very striking. Not only do the physical agents, in their actions, all lead to one result, viz., an accumulation of snow and ice when the winters are in aphelion, but their efficiency in bringing about this result is actually strengthened by their mutual reactions. To illustrate this effect I may quote the following from Chapter III. of this volume.

'To begin with, we have a high state of eccentricity. This leads to long and cold winters. The cold leads to snow, and although heat is given out in the formation of the snow, yet the final result is that the snow intensifies the cold: it cools the air and leads to still more snow. The cold and snow bring a third agent into play—*fogs*, which act still in the same direction. The fogs intercept the sun's rays; this interception of the rays diminishes the melting-power of the sun, and so increases the accumulation. As the snow and ice continue to accumulate, more and more of the rays are cut off; and on the other hand, as the rays continue to be cut off, the *rate* of accumulation increases, because the quantity of snow and ice melted becomes thus annually less and less. In addition, the loss of the rays cut off by the fogs lowers the temperature of the air and leads to more snow being formed, while again the snow thus formed chills the air still more and increases the fogs. Again, during the winters of a glacial epoch, the earth would be radiating its heat into space. Had this loss of heat simply lowered the temperature, the lowering of the temperature would have tended to diminish the rate of loss; but the result is the formation of snow rather than the lowering of the temperature.

'Further, as snow and ice accumulate on the one hemisphere they diminish on the other. This increases the strength of the trade-winds on the cold hemisphere and weakens those on the warm. The effect of this is to impel the warm water of the tropics more to the warm hemisphere than to the cold. Supposing the northern hemisphere to be the cold one, then, as the snow and ice begin gradually to accumulate, the ocean-currents of that hemisphere, more particularly the Gulf Stream, begin to decrease in volume, while those on the southern or warm hemisphere begin *pari passu* to increase. This withdrawal of heat from the northern hemisphere favours the accumulation of snow and ice, and as the snow and ice accumulate the ocean-currents decrease. On the other hand, as the ocean-currents diminish, the snow and ice still more accumulate. Thus the two effects, in so far as the accumulation of snow and ice is concerned, mutually strengthen each other.'

With all this Mr. Wallace seems fully to agree; for at pp. 137–140 ("Island Life") he gives a very clear statement of the effect of these mutual reactions in the production of glaciation, and says that were it not for them it is probable the astronomical and other causes would not in our latitudes have been sufficient to produce glaciation. In short, he concludes that these reactions "produce a maximum of effect which, without their aid, would be altogether unattainable." Mr. Wallace thus does full justice to these mutual reactions in so far as the production of glaciation is concerned; but I am convinced that he must have under-estimated their importance as regards the removal of the glaciation. He, however, recognises the fact that these mutual reactions produce an opposite effect on the warm hemisphere whose winters are in perihelion. "These agencies," he says, "are at the same time acting in a reverse way in the southern hemisphere, diminishing the supply of the moisture carried by the anti-trades, and increasing the temperature by means of more powerful southward ocean-currents; and all this again reacts on the northern hemisphere, increasing yet further the supply of moisture by the more powerful south-westerly winds, while still further lowering the temperature by the southward diversion of the Gulf Stream."

Now, if, during the glaciation of the northern hemisphere, these mutual relations produce the opposite effect on the southern hemisphere, it is evident that they must produce this same opposite effect on the northern hemisphere when its winter solstice is in perihelion. Their effect then would be to increase the temperature and melt the ice. When the winter solstice is moving towards the aphelion, the physical agents begin to act and react on one another, and this action and reaction go on increasing in intensity till the solstice-point reaches the aphelion; but an exactly similar thing is going on in the other hemisphere, only the effects are the reverse. While the actions and reactions leading to an accumulation of ice are increasing in intensity, we shall suppose, on the northern hemisphere, the same increase is taking place on the southern hemisphere; but the result is a melting, not an accumulation of the ice. The same process is undoing on the southern hemisphere what it is doing on the northern. Similarly, of course, when the northern winter solstice begins to move towards the perihelion, the mutual reactions of these physical causes will be reversed and will go on with increasing intensity till the perihelion is reached, melting the very ice which they had previously produced.

We have already seen that the greater the extent of the eccentricity the greater is the effect resulting from the *actions* of the physical causes, whether this effect be the production of ice on the cold hemisphere, or its removal from the warm. It is evident that the same thing must necessarily hold true in regard to the mutual *reactions* of the physical causes. Consequently, if the mutual actions and reactions of the physical causes, brought into operation during a high state of eccentricity, led at the Glacial Epoch to the great accumulation of ice when the winters were in aphelion, they must have led to an equally great melting and dispersal of that ice when precession brought the winters round to perihelion. These causes would be as efficient in the removal of the ice as they were in its production. In so far as the physical and astronomical causes were concerned, the greater the amount of ice formed during the cold periods the greater would be the amount melted during the warm interglacial periods.

This conclusion follows so obviously from the foregoing principles that it seems almost superfluous to dwell further on the subject, the more so as it will be seen in the next chapter that the correctness of the conclusion is established by the facts of geology and palæontology.

Another Reason Assigned why the Ice does not Melt.—Mr. Wallace assigns the following as an additional reason why the ice does not disappear during the interglacial periods when the eccentricity is high:—

"When a country is largely covered with ice, we may look upon it as possessing the accumulated or stored-up cold of a long series of preceding winters; and however much heat is poured upon it, its temperature cannot be raised above the freezing-point till that store of cold is got rid of—that is, till the ice is all melted. But the ice itself, when extensive, tends to its own preservation, even under the influence of heat; for the chilled atmosphere becomes filled with fog, and this keeps off the sun-heat, and then snow falls even during summer, and the stored-up cold does not diminish during the year. When, however, only a small portion of the surface is covered with ice, the exposed earth becomes heated by the hot sun; this warms the air, and the warm air melts the adjacent ice. It follows that, towards the equatorial limits of a glaciated country alternations of climate may occur during a period of high eccentricity, while nearer the Pole, where the whole country is completely ice-clad, no amelioration may take place." *

For the past twenty years I have been maintaining that, when a country is covered with ice, it becomes a permanent source of cold; and however much heat may be received from the sun, the temperature of the surface can never be raised above the freezing-point while the ice remains; and, again, that such an ice-covering tends to its own preservation, because it chills the air and increases the snowfall. In short, I have all along maintained this to have been one of the

* "Island Life," p. 154.

chief causes which led to the country being so deeply covered with ice. In fact, had it not been for some such conservative power in the ice, a glacial epoch resulting from the causes which I have been advocating would not have been possible. This conservative tendency certainly renders it more difficult for the physical agencies to get rid of the ice during inter-glacial periods; but we evidently have no grounds for assuming that it will defy their melting-powers.

CHAPTER VIII.

EXAMINATION OF MR. ALFRED R. WALLACE'S MODIFICATION OF THE PHYSICAL THEORY OF SECULAR CHANGES OF CLIMATE—*Continued.*

Professor J. Geikie on Condition of Europe during Interglacial Periods.—Scotland during Interglacial Periods.—Difficulty in Detecting the Climatic Character of the Earlier Interglacial Periods.—Objection as to the Number of Interglacial Periods.—Objection as to the Number of Submergences.—Interglacial Periods less strongly marked in Temperate Regions than Glacial.

II. *Geological and Palæontological Facts in relation to Mr. Wallace's Modification of the Theory.*

Mr. Wallace's chief, and indeed only real, modification of my theory, is to the effect, as I have pointed out, that the alternate phases of precession, causing the winter of each hemisphere to be in aphelion and perihelion each 10,500 years, would produce a complete change of climate only when a country was partially snow-clad. According to his view, when the greater part of North-western Europe was almost wholly buried under snow and ice, those glacial conditions must have continued, and perhaps have even become intensified, when the winter solstice moved round to perihelion, instead of being replaced, as I have maintained, by an almost perpetual spring. In short, Mr. Wallace's conclusion is that, during the Glacial Epoch proper, a warm and equable Interglacial Period could not have occurred.

In the preceding chapter, I have endeavoured to show that physical principles do not warrant such a conclusion. I shall now proceed to consider what the direct testimony of Geology and Palæontology is on the subject; and I believe we shall find that the facts of Geology and Palæontology are as much opposed to the conclusion as are the principles of Physics.

On this point I may quote the evidence of a geologist who, more than any other, has devoted special attention to all points relating to Glacial and Interglacial periods. Prof. J. Geikie, after devoting upwards of five hundred pages of his "Prehistoric Europe" to the consideration and accumulation of facts from all parts of this country and the Continent relating to Glacial and Interglacial periods, gives the following as the result of his investigations:—

"We note," he says, "as we advance from Pliocene times, how the climatic conditions of the colder epochs of the Glacial Period increase in severity until they culminate with the appearance of that great northern *mer de glace* which overwhelmed all Northern Europe, and reached as far south as the 50th parallel of latitude in Saxony. Thereafter the glacial epochs decline in importance, until, in the Postglacial Period, they cease to return. The genial climate of Interglacial ages probably *also attained a maximum* towards the *middle* of the Pleistocene Period, and afterwards became less genial at successive stages, the temperate and equable conditions of early Postglacial times being probably the latest manifestation of the Interglacial phase." *

I shall now quote the same author's description of an Interglacial Period as demonstrated by its flora and fauna. The reader must, however, observe that,

* "Prehistoric Europe," p. 561.

by Pleistocene Period, Professor Geikie means the so-called Glacial Period, with its alternations of severe Arctic climate and mild and genial conditions. *

"An examination," he says, "of Pleistocene organic remains leads us to conclude that strongly-contrasted climatic conditions alternated during the Period. At one time an extremely equable and genial climate prevailed, allowing animals, which are now relegated to widely-separated zones, to live throughout the year in one and the same latitude. Hippopotamuses, elephants, and rhinoceroses, Irish deer, horses, oxen, and bisons then ranged from the borders of the Mediterranean as far north at least as Middle England and Northern Germany. In like manner, plants which no longer occur together—some being banished to hilly regions, while others are restricted to low grounds, and yet others have retreated to the extreme south of the Continent or to warmer regions beyond the limits of Europe—lived side by side. The fig-tree, the judas-tree, and the Canary laurel flourished in Northern France along with the sycamore, the hazel, and the willow. And we encounter in the Pleistocene deposits of various countries in Europe the same remarkable commingling of northern and southern forms—of forms that demand a humid climate and are capable of enduring considerable cold, together with species which, while seeking moist conditions, yet could not survive the cold of our present winters. The testimony of the mammals and plants is confirmed by that of the land and fresh-water mollusca—all the evidence thus conspiring to demonstrate that the climate of Pleistocene Europe was, for some time at all events, remarkably equable and somewhat humid.

* "Prehistoric Europe," p. 544.

The summers may not indeed have been warmer than they are now; the winters, however, were certainly much more genial." *

This, be it observed, is a description of a condition of things which existed during an Interglacial Period belonging, not to the close, but to the very climax of the Glacial Epoch. For, immediately preceding and succeeding this Period, almost the whole of Northern Europe was enveloped in one continuous sheet of ice. "But if," continues Professor J. Geikie, "the evidence of such a climate having formerly obtained be very weighty, not less convincing are the proofs, supplied by the Pleistocene deposits, of extreme conditions. Think what must have been the state of Middle and Northern Europe when Palæolithic man hunted the reindeer in Southern France, and when the Arctic willow and its congeners grew at low levels in Central Europe. Reflect upon the fact that in the very same latitude in France, where at one time the Canary laurel and the fig-tree flourished, the pine, the spruce, and northern and high-alpine mosses at another time found a congenial habitat. Bear in view, also, that the land and fresh-water molluscs testify in like manner to the same strongly-contrasted climate. Besides those that tell of more equable and genial conditions than the present, there are species now restricted to the higher Alps and northern latitudes that formerly abounded in Middle Europe, and their shells occur commingled in the same deposits with the remains of lemmings, marmots, reindeer, and other northern and mountain-loving animals." †

But more convincing still is another range of facts, some of which have been adduced by Mr. Wallace

* "Prehistoric Europe," p. 540.

† Ibid, p. 541.

himself. In a section on alternations of warm and cold periods during the Glacial Epoch,* he says:—

"The evidence that such was the case" (alternate warm and cold periods) "is very remarkable. The 'Till,' as we have seen, could only have been formed when the country was entirely buried under a large ice-sheet of enormous thickness, and when it must therefore have been, in all the parts so covered, almost entirely destitute of animal and vegetable life. But in several places in Scotland fine layers of sand and gravel, with beds of peaty matter, have been found resting on 'Till,' and again covered by 'Till.' Sometimes these intercalated beds are very thin, but in other cases they are twenty or thirty feet thick, and in them have been found remains of the extinct ox, the Irish elk, the horse, reindeer, and mammoth. Here we have evidence of two distinct periods of intense cold, and an intervening milder period sufficiently prolonged for the country to become covered with vegetation and stocked with animal life."

Let us now see to what all this leads. It has been proved beyond the possibility of a doubt that, at the time the Till was being formed which *overlies* the Scottish interglacial beds, the whole of Scotland, Scandinavia, the bed of the North Sea, and a great part of the North of England were covered with one continuous sheet of ice upwards of 2000 feet in thickness. This sheet overwhelmed the Hebrides, the Orkney and Shetland Islands, extended into Russia, filled the basin of the Baltic, overflowed Denmark and Holstein, and advanced into North Germany as far at least as Berlin. It has also been demonstrated that, at the time the Lower Till was being formed which *underlies* these interglacial beds, North-western

* "Island Life," p. 114.

Europe was under a still more severe state of glaciation. The ice-sheet at this time advanced farther south into England, and extended into North Germany as far as Saxony. It is perfectly obvious that this sheet must have destroyed all plant and animal life in Scotland; and before the country could have become covered with vegetation and stocked with those interglacial animals to which Mr. Wallace refers, the ice must have disappeared and the climate become mild.

Equally conclusive are the facts adduced by Mr. Wallace in reference to the interglacial beds of England. "In the east of England, Mr. Skertchly," he says, "enumerates four distinct boulder-clays with intervening deposits of gravels and sands. Mr. Searles V. Wood, jun., classes the most recent (Hessle) boulder-clay as 'Postglacial,' but he admits an intervening warmer period, characterised by southern forms of mollusca and insects, after which glacial conditions again prevailed with northern types of mollusca. Elsewhere Mr. Wood says:—'Looking at the presence of such fluviatile mollusca as *Cyrena fluminalis* and *Unio littoralis,* and of such mammalia as the hippopotamus and other great pachyderms, and of such a littoral Lusitanian fauna as that of the Selsea bed, where it is mixed up with the remains of some of those pachyderms, as well as of some other features, it has seemed to me that the climate of the earlier part of the Postglacial Period in England was possibly even warmer than our present climate; and that it was succeeded by a refrigeration sufficiently severe to cause ice to form all round our coasts, and glaciers to accumulate in the valleys of the mountain districts.'" That these fauna indicate a warm and equable condition of climate is further evident from Mr. Wallace's remarks:—"The fact," he says, "of the hippopotamus having lived at 54° N. lat. in England, quite close to the time of the Glacial Epoch, is absolutely inconsistent with a mere gradual amelioration of climate from that time till the present day. The immense quantity of vegetable food which this creature requires, implies a mild and uniform climate with hardly any severe winter; and no theory that has yet been suggested renders this possible, except that of alternate cold and warm periods during the Glacial Epoch itself. . . . Thus the very existence of the hippopotamus in Yorkshire, as well as in the south of England, in close association with glacial conditions, must be held to be a strong corroborative argument in favour of the reality of an interglacial warm period."

I trust that Mr. Wallace has not been misled by Mr. Wood's unfortunate use of the term "Postglacial" as applied to the Hessle boulder-clay. The Hessle boulder-clay as surely belongs to the Glacial Period proper as does the true Till of Scotland, which covers the Lowlands and overlies the interglacial beds of that country. It is the *moraine profonde* of the last *mer de glace*, which covered the greater part of North-western Europe. The Upper Till of Scotland and the Hessle boulder-clay of England belong to the same period. This has been clearly shown by Professor J. Geikie in his "Great Ice-Age," chap. xxx. (2nd edit.), and in "Prehistoric Europe," chap. xii., and elsewhere. The Hessle boulder-clay is, in short, a continuation of the Upper Till of Scotland.

The position of these Hessle beds to which Mr. Wallace refers, like that of the interglacial beds of Scotland, is between two boulder-clays—the Hessle and the Purple boulder-clays, both of which indicate a period of extreme glaciation: only the Purple boulder-clay period was somewhat the more severe

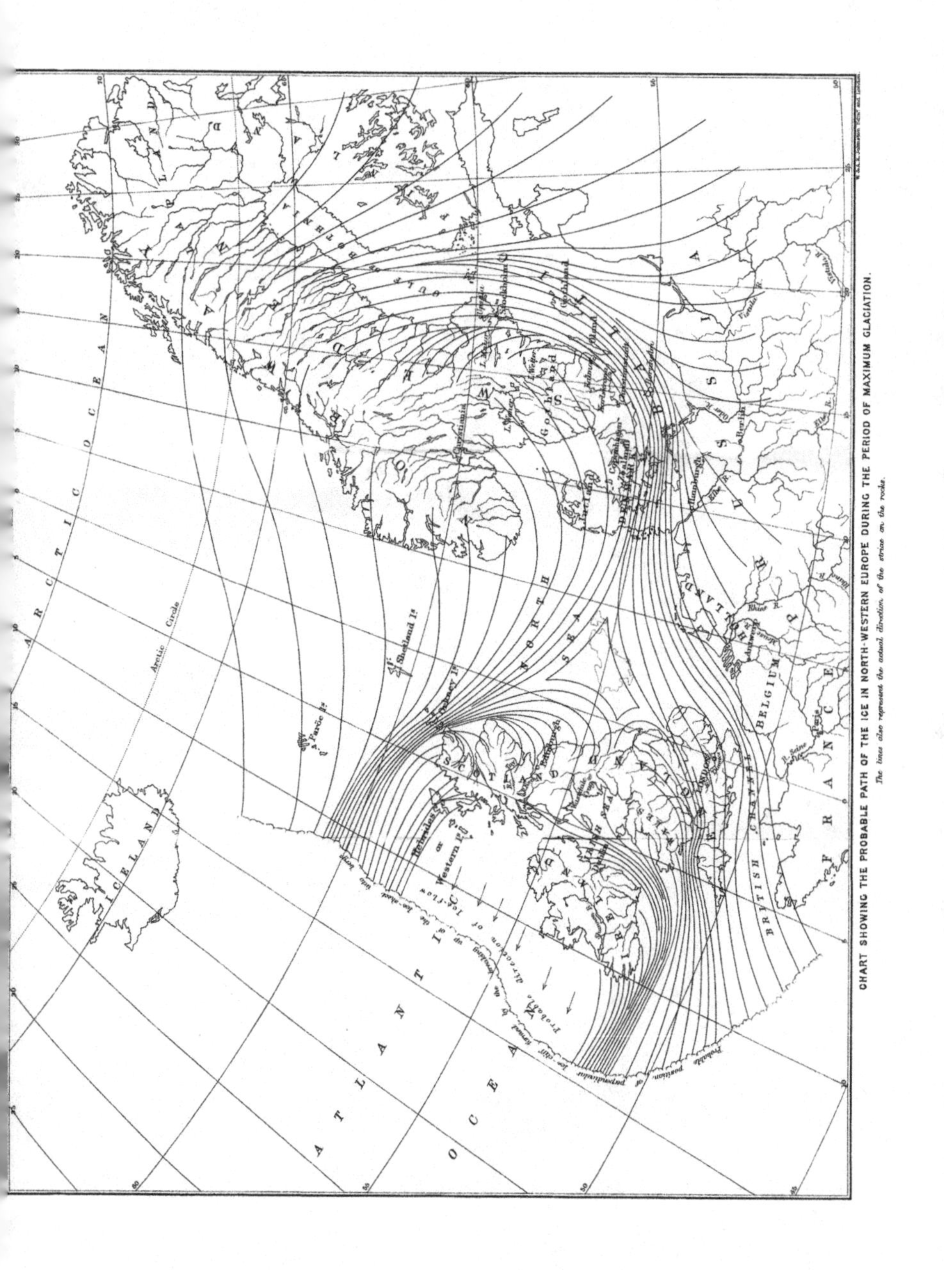

CHART SHOWING THE PROBABLE PATH OF THE ICE IN NORTH-WESTERN EUROPE DURING THE PERIOD OF MAXIMUM GLACIATION.

The lines also represent the actual direction of the striae on the rocks.

The material originally positioned here is too large for reproduction in this reissue. A PDF can be downloaded from the web address given on page iv of this book, by clicking on 'Resources Available'.

of the two. At both periods, the greater part of North-western Europe was buried under ice. We know that during the last great ice-period, which was undoubtedly the period of the Hessle boulder-clay, the ice-sheet reached in North Germany as far as Berlin; while during the period of the Purple boulder-clay it advanced to about Saxony.

The accompanying chart, reproduced from 'Climate and Time,' which was sketched out during the summer of 1870, shows pretty correctly the condition of North-western Europe both before and after the interglacial period referred to by Mr. Wood. The observations of Professor J. Geikie and Mr. A. Helland have since shown, however, that the Scandinavian land-ice did not pass over the Faröe Islands, as represented in the chart; but the chart has, in almost every other particular, been now proved by geologists to be accurate. The chart exhibits in a striking manner the enormous amount of ice which must have been melted off the ground before the warmth of the interglacial period could even have commenced.

The observations of Prof. Torrell, Dr. A. Penck, Prof. Credner, Prof. Berendt, Dr. Jentzsch, A. Helland, F. Wahnschaffe, H. Habenicht, and other geologists, have shown that there are in North Germany three distinct boulder-clays—an Upper, Middle, and Lower, with two series of interglacial beds. In these interglacial beds have been found organic remains which evidently indicate a mild and genial condition of climate. The younger interglacial period (the one prior to the last great extension of the ice) in all probability corresponds to the last interglacial period of Scotland, England, and Ireland. Interglacial beds belonging to the same period have been found in Switzerland, Italy, Denmark, North America, and other places, all indicating a mild and equable condition of climate.

There is another class of facts, almost entirely overlooked, which will doubtless yet prove as conclusively the warm character of interglacial periods. These facts will be referred to when we come to consider the question of warm polar climates.

It would be impossible at present to give even the briefest outline of the recent discoveries in regard to interglacial periods. But though this were possible it would be wholly unnecessary, as the facts which have already been adduced by Mr. Wallace himself are perfectly sufficient for our present purpose.

If now it be true, as it undoubtedly is, that the Hessle boulder-clay of England belongs to the same age as the Upper Till of Scotland, and that the last warm interglacial period—when the *Cyrena fluminalis* and *Unio littoralis*, the hippopotamus, the *Elephas antiquus*, and other animals of a southern type lived in England—occurred between two glacial periods so severe as to envelop the greater part of North-western Europe in a continuous sheet of ice, then this particular interglacial period must have supervened during a high state of eccentricity, and not, as Mr. Wallace assumes, at a period subsequent to the Glacial Epoch proper, when the eccentricity had greatly diminished. This is obvious; for if the last great ice-sheet could have been produced without a high state of eccentricity, then there seems no reason why the one preceding it should not also have been produced without high eccentricity. If so, then all the previous ice-sheets may in like manner have been so produced. For the difference in magnitude between the last and penultimate ice-sheets was not so great as to warrant the supposition of any considerable difference in the amount of eccentricity at the two periods when these ice-sheets were respectively developed. In short, if the last great ice-sheet can be explained without the supposition of a high state of eccentricity, then there does not appear to be any real necessity for any theory of eccentricity in accounting for the Glacial Epoch.

If we adopt the Physical theory of the cause of the Glacial Epoch, we are compelled to maintain that the last two great Ice-periods were the indirect results of a high state of eccentricity, and in this case we can hardly avoid the conclusion that the mild intervening period was due to the same cause. The occurrence of a mild interglacial period between the two ice-periods is directly in opposition to Mr. Wallace's view—that during a high state of eccentricity the ice would not disappear but be continued. It is in perfect harmony, however, with that which I advocate; for during high eccentricity a mild and equable condition of climate, when the winters occur in perihelion, is as much a necessary result as a cold and glacial condition when they occur in aphelion.

The facts of Geology thus to me appear, so far, to be as much opposed to Mr. Wallace's modifications as are the principles of Physics.

Difficulty in detecting the Climatic Character of the Earlier Interglacial Periods.—It follows according to theory that, other things being equal, the greater the amount of eccentricity the more equable and mild will the interglacial periods be. It is probable, therefore, that some of the earlier interglacial periods were milder and more equable than the last. It may be difficult, in the present state of our knowledge, to prove this conclusion by direct geological and palæontological evidence; but, on the other hand, it is certainly impossible to disprove it by that means. The absence of deposits containing organic remains, indicative of a superior mildness of climate having obtained during early interglacial periods, cannot certainly be regarded as satisfactory evidence against the conclusion just referred to. When we consider the enormous pressure and destructive power of an ice-sheet some 2000 or 3000 feet in thickness grinding down the face of a country, our surprise is that so much evidence remains of even the last interglacial period. That so few relics of the flora and fauna of preceding interglacial periods have been preserved, is a conclusion which we might *à priori* anticipate. This fact has been clearly pointed out by Mr. Wallace himself, who says:—"If there have been, not two only, but a series of such alternations of climate, we could not possibly expect to find more than the most slender indications of them, because each succeeding ice-sheet would necessarily grind down or otherwise destroy much of the superficial deposits left by its predecessors, while the torrents that must always have accompanied the melting of these huge masses of ice would wash away even such fragments as might have escaped the ice itself." *

When we pass beyond the limits reached by the ice-sheets of the Glacial Epoch, we may expect, of course, to find the remains of many of the plants and animals which lived during the earlier interglacial periods. But here, again, we encounter another difficulty; for we have in this case seldom any means of determining the age to which these remains belong. Unless in relation to overlying and underlying boulder-clays, there seems in many cases no way of knowing to what interglacial period they ought to be assigned; or, in fact, whether they are really interglacial or not. If the remains in question indicated a condition of

* "Island Life," p. 118.

climate much milder than the present, the probability is that they would be classified as preglacial. I fully agree with Prof. J. Geikie, that many of those plants and animals of a southern type which have been regarded as preglacial are in reality of interglacial age.

Objection as to the Number of Interglacial Periods.—It has been urged as an objection to the Physical theory of the Glacial Epoch, that, according to it, there ought to have been more interglacial periods than we have direct evidence of having actually occurred. I am doubtful as to the force of this objection. I do not think that there could have been more than about five well-marked interglacial periods during the entire Glacial Epoch; three probably during the former half of the epoch, and certainly not more than two during the latter half. There would be a large interval between the two maxima of eccentricity of 100,000 and 200,000 years ago, when the alternations of climate would be comparatively moderate in extent. Besides, it is not correct to assume, as is generally done, that the interval between two consecutive interglacial periods is only 21,000 years; for the mean rate of motion of the perihelion during the Glacial Epoch was considerably less than has been assumed. It will be seen from the Table of the Longitude of the Perihelion, given in 'Climate and Time,' p. 320, that it has taken the perihelion 231,000 years to make one complete revolution. If, therefore, we assume, what of course is not certain, that the mean rate of precession during the Glacial Epoch was the same as the present, then the rate of precession to that of the perihelion's motion would, in this case, be as nine to one. The equinoxial point will take 25,811 years to make one revolution; but as the perihelion moves in the opposite direction, it will reduce the time taken by the point in passing from perihelion round to perihelion to 23,230 years, which will represent the mean interval between two consecutive interglacial periods. But as the motion of the perihelion was very irregular, the length of the interval between the periods would, of course, differ considerably.

When we consider how difficult it must be to detect in the drift covering glaciated countries even a relic of early interglacial deposits, and when, moreover, we remember that it is only within the past few years that geologists have begun to bestow any attention on the subject, it is certainly not surprising that direct geological evidence of so few interglacial periods has as yet been discovered. In England, geologists have, however, already detected evidence of three interglacial periods, with four or five ice-periods. In Germany, quite recently, two interglacial periods and three or more ice-periods have been recognised by competent observers. In Denmark there are four boulder-clays separated by intercalated beds of sand and clay. In severely-glaciated Scotland, where traces of former interglacial periods can hardly be expected, there have nevertheless been found in old preglacial buried channels, and other sheltered hollows, three, four, and in some places five, boulder-clays, separated from one another by immense beds of sand, gravel, and clay. Some of these beds are found to be continuous for long distances. It is true that these intercalated beds have yielded few or no organic remains, but it may well be that further research will yet result in the discovery of more abundant fossils; for frequently the beds in question are too thick and too extensive to allow us to infer their subglacial origin. They do not in such respects resemble the deposits which have been accumulated by aqueous action under ice, but have all the characteristics of deposits which have been laid down in lakes and lacustrine hollows. As some have already yielded organic remains, a more extended scrutiny will probably lead to the discovery of similar fossils in those beds which are at present believed to be unfossiliferous.

Objection as to the number of Submergences.—It has also been urged as an objection to the Physical theory, that, according to it, there ought to have been a greater number of submergences of the land during the glacial epoch than is known to have taken place; for according to the theory there ought to have been a submergence to a greater or lesser extent corresponding to each ice period. Submergence ought regularly to accompany glaciation, and emergence the disappearance of the ice; whereas geologists have detected only about three such periods.

In reply to this objection, it will not do to estimate the number of submergences by the number of observed raised beaches or well-marked terraces. These in most cases must have resulted from a subsidence of the land, and not from a rise of the sea due to a displacement of the earth's centre of gravity. Oscillations of sea level resulting from an alternate increase and decrease of ice would not likely produce well-marked terraces. In order to cut a terrace the sea must continue for a long period at the same level; but this could hardly be expected in the case of a rise resulting from the accumulation of ice; for, according to theory, the mass of the ice gradually increases till a maximum is reached, when it then begins as gradually to decrease. It is true that when the ice is near its maximum it will change but slowly, and when at the turning point it may remain stationary for centuries before any sensible decrease takes place. In this case it might in some places leave its mark in the form of a terrace or a raised beach; but no doubt, such cases would be exceptional. The condition of things becomes further complicated by another cause referred to in 'Climate and Time' (p. 388), which will occasionally come into operation; viz., a lowering of the general level of the ocean resulting from the abstraction of the ice, or a rise resulting from a general decrease of the ice.

The submergences and emergences arising from displacement of the earth's centre of gravity would of course leave evidence of their existence in the form of stratified deposits; but, as we have already seen, no one could possibly determine from such deposits the number of elevations or depressions of sea level which actually took place.

I think it is probable, however, that some of the more recent well-marked changes of sea level, such as those indicated by the Carse-clays and the submarine Forest-beds, were due to displacements of the earth's centre of gravity. I am inclined also to believe that the rise of the land, or rather the lowering of the sea level during the interglacial or continental periods, in many cases, resulted from the same cause. If we admit, with some geologists, that the sinking of the land was due to the weight of the ice, we shall have an explanation of glacial submergences; but such a theory will in no wise explain the elevation of the land during the continental periods. It is true, the removal of the ice might allow the land to regain its former level; but its removal could have no tendency to raise the land above that level.

The whole matter of glacial submergence is too obscure and complicated an affair to allow us to determine, with anything like certainty, how often the land might have been under water during the glacial epoch.

Interglacial Periods Less Strongly Marked in Temperate Regions than Glacial.—I quite agree with Mr. Wallace that the interglacial deposits never exhibit any indication of a climate whose warmth corresponded to the severity of the preceding cold. This, however, cannot be urged as an objection, for it is a result which follows as a necessary consequence from theory. It theoretically follows that the cold of the glacial periods will not only exceed in severity the heat of the interglacial, but will also be of longer duration. During the glacial periods extreme cold is the characteristic of the winters, which, owing to the presence of snow and ice, only becomes moderated, although, of course, considerably, during the summers. But, on the other hand, during interglacial periods mildness and equability of temperature rather than heat are the characteristics both of summer and winter.

That the cold of the glacial periods must have continued longer than the warmth of the interglacial will, I think, be apparent from the following considerations. As long as a country remains permanently covered with snow and ice, the climate, as has been repeatedly shown, must continue cold, no matter what the direct heat of the sun may be. Astronomically considered, the interglacial periods are, of course, of the same length as the glacial,—the mean length of which, during the Glacial Epoch, was about 11,600 years; but the cold of a glacial period would not, as we shall presently see, actually terminate at the end of the period, but would be continued on probably for centuries into the succeeding interglacial period. Suppose that during a glacial period the country is covered with a sheet of ice, which, during the continuance of the period, had accumulated to the thickness of 2000 or 3000 feet. All this enormous quantity of ice would have to be melted off the ground before the warmth of the interglacial period would commence. So long as a single inch of ice covered the surface of the country, the cold would continue. Ice, as we have seen, by chilling the air, induces fresh snow to fall; and, of course, it is only when the amount of ice annually melted exceeds that being formed from the falling snow, that a diminution in the thickness of the sheet would begin to take place. A real melting of the ice, and consequent decrease in the thickness of the sheet, would probably not commence till the astronomical and physical agencies in operation during the glacial period began to act in an opposite direction. In short, it would be the favourable conditions of the interglacial period that would effectually remove the ice; and it would be then, and only then, that the warmth would begin; while, again, at the close of the period, when the first inch of ice made its appearance on the surface of the country, the interglacial condition of climate would come to an end. The time required to remove the ice does not prevent an interglacial condition of climate; it only somewhat shortens its duration.

There is another circumstance worthy of notice here. It is this: as the mild and equable character of the climate during interglacial periods resulted to a large extent from the enormous transference of equatorial heat, and its distribution over temperate and polar regions, the difference of climatic conditions between the subtropical and the temperate and polar regions would be less marked than at present; in other words, the temperature would not differ so much with latitude as it does at present. This, as we have seen, is a conclusion which is fully borne out by geological and palæontological facts.

CHAPTER IX.

THE PHYSICAL CAUSE OF MILD POLAR CLIMATES.

The Probably True Explanation.—Sir William Thomson on Mild Arctic Climates.—Mr. Alfred R. Wallace on Mild Arctic Climates. — Influence of Eccentricity during the Tertiary Period.

THERE are few facts within the domain of geology better established than that at frequent periods in the past the polar regions enjoyed a comparatively mild and equable climate, and that places now buried under permanent snow and ice were then covered with a rich and luxuriant vegetation. Various theories have been advanced to account for this remarkable state of things, such as a different distribution of sea and land, a change in the obliquity of the ecliptic, a displacement in the position of the earth's axis of rotation, and so forth. The true explanation will, I feel persuaded, be found to be the one I gave many years ago. The steps by which my conclusions were reached were as follows:—

The annual quantity of heat received from the sun at the equator is to that at the poles as 12 to 4·98, or, say, as 12 to 5. This is on the supposition that the same percentage of rays is cut off by the atmosphere at the equator as at the poles, which, of course, is not the case. More is cut off at the poles than at the equator, and consequently the difference in the amount of heat received at the two places is actually greater than that indicated by the ratio 12 to 5. But, assuming 12 to 5 to be the ratio, the question arose what ought to be the *difference* of temperature between the two places in question on the supposition that the temperature was due solely to the direct heat received from the sun? This was a question difficult to answer, for its answer mainly depended upon two considerations, regarding both of which a very considerable amount of uncertainty prevailed.

First, it was necessary to know how much of the total amount of heat received by the earth was derived from the sun, and how much from the stars and other sources, or, in other words, from space. Absolute zero is considered to be $-461°$ Fahr. The temperature of the equator is about 80°. This gives 541° Fahr. as the absolute temperature of the equator. Now, were all the heat received by the earth derived simply from the sun, and were the temperature of each place proportionate to the amount directly received, then the absolute temperature of the poles would be $\frac{5}{12}$ of that of the equator, or 225°. This would give a difference of 316° between the temperature of the equator and that of the poles. According to Pouillet and Herschel, space has a temperature of $-239°$, or 222° of absolute temperature. If this be the temperature of space, then only 319° of the absolute temperature of the equator are derived from the sun; consequently, as the poles receive from the sun only $\frac{5}{12}$ of this amount of temperature, or 133°, this will give merely 186° as the difference which ought to exist between the equator and the poles. There is, however, good reason for believing that the temperature of space is far less than that assigned by Pouillet and Herschel —that, in fact, it is probably not far above absolute zero. Therefore, by adopting so high a temperature as – 239°, we make the difference between the temperature of the equator and that of the poles too small.

Second, it was necessary to know at what rate the temperature increased or decreased with a given increase or decrease in the amount of heat received. It was well known that Newton's law—that the change of temperature was directly proportionate to the change in the quantity of heat received—was far from being correct. The formula of Dulong and Petit was found to give results pretty accurate within ordinary limits of temperature. But it would not have done, in making my estimate, to take that formula, if I adopted Herschel's estimate of the temperature of space; for it would have made the difference of temperature between the equator and the poles by far too small. Newton's law, if we adopt Herschel's estimate of the temperature of space, would give results much nearer the truth; for the error of the one would, to a large extent at least, neutralise that of the other.

From such uncertain data it was, of course, impossible to arrive at results which could in any way be regarded as accurate. But it so happens that perfect accuracy of results in the present case was not essential; all that really was required was a rough estimate of what the difference of temperature between the equator and the poles ought to be. The method adopted showed pretty clearly, however, that the difference of temperature could not be less (although probably more) than 200°; but the present actual difference does not probably exceed 80°. We have no means of ascertaining with certainty what the mean annual temperature of the poles is; but as the temperature of lat. 80° N. is 4°·5, that of the poles is probably not under 0°. If the present difference be

80°, it is then 120° less than it would be did the temperature of each place depend alone on the heat received directly from the sun. This great reduction from about 200° to 80° can, of course, be due to no other cause than to a transference of heat from the equator to the poles. The question then arose, by what means was this transference effected? There were only two agencies available—the transference must be effected either by aerial or by ocean-currents. It was shown at considerable length ('Climate and Time,' pp. 27–30, and other places) that the amount of heat that can be conveyed from the equator to the poles by means of aerial currents is trifling, and that, consequently, the transference must be referred to the currents of the ocean. It became obvious then that the influence of ocean-currents in the distribution of heat over the globe had been enormously underestimated. In order to ascertain with greater certainty that such had been the case, I resolved on determining, if possible, in absolute measure, the amount of heat actually being conveyed from the equator to temperate and polar regions by means of ocean-currents.

The only great current whose volume and temperature had been ascertained with any degree of certainty is the Gulf Stream. On computing the absolute amount of heat conveyed by that stream, it was found to be more than equal to all the heat received from the sun within 32 miles on each side of the equator. The amount of equatorial heat carried into temperate and polar regions by this stream alone is therefore equal to one-fourth of all the heat received from the sun by the North Atlantic from the Tropic of Cancer up to the Arctic circle.* Although the heating-

* 'Climate and Time,' pp. 34, 35; "Phil. Mag.," February, 1870.

power of the Gulf Stream had long been known, yet no one had imagined that the warmth of our climate was due, to such an enormous extent, to the heat conveyed by that stream. The amount of heat received by an equatorial zone 64 miles in breadth represents, be it observed, merely the amount conveyed by one current alone. There are several other great currents, some of which convey as much heat polewards as the Gulf Stream. On taking into account the influence of the whole system of oceanic circulation, it is not surprising that the difference of temperature between the equator and the poles should be reduced from 200° to 80°.

From these considerations, the real cause of former comparatively mild climates in Arctic regions becomes now apparent. All that was necessary to confer on, say, Greenland a condition of climate which would admit of the growth of a luxuriant vegetation is simply an increase in the amount of heat transferred from equatorial to Arctic regions by means of ocean-currents. And to effect this change of climate no very great amount of increase is really required; for it was shown that the severity of the climate of that region is about as much due to the cooling effect of the permanent snow and ice as to an actual want of heat. An increase in the amount of warm water entering the Arctic Ocean, just sufficient to prevent the formation of permanent ice, is all that is really necessary; for were it not for the presence of ice the summers of Greenland would be as warm as those of England.

Were the whole of the warm water of the Gulf Stream at present to flow into the Arctic Ocean, it would probably remove the ice of Greenland. Any physical changes, such as those that have been discussed on former occasions, which would greatly increase the volume and temperature of the stream and deflect more of its waters into the Arctic Ocean would, there is little doubt, confer on the polar regions a climate suitable for plant and animal life. At present the Gulf Stream bifurcates in mid-Atlantic, one branch passing north-eastwards into the Arctic regions, whilst the larger branch turns south-eastwards by the Azores, and after passing the Canaries re-enters the equatorial current. As the Gulf Stream, like other great currents of the ocean, follows almost exactly the path of the prevailing winds *, it bifurcates in mid-Atlantic simply because the winds blowing over it bifurcate also. Any physical change which would prevent this bifurcation of the winds and cause them to blow north-eastwards would probably impel the whole of the Gulf Stream waters into the Arctic seas. All this doubtless might quite well be effected without any geographical changes, although changes in the physical geography of the North Atlantic might be helpful.

These considerations regarding the influence of the Gulf Stream point to another result of an opposite character. It is this: if a large *increase* in the volume and temperature of the stream would confer on Greenland and the Arctic regions a condition of climate somewhat like that of North-western Europe, it is obvious, as has been shown at length on former occasions, that a large *decrease* in its temperature and volume would, on the other hand, lead to a state of things in North-western Europe approaching to that which now prevails in Greenland. A decrease leads to a glacial, an increase to an interglacial condition of things.

Sir William Thomson on Mild Arctic Climates.—

* See 'Climate and Time,' p. 213.

In a paper read before the Geological Society of Glasgow in February, 1877, Sir William maintains also that an increase in the amount of heat conveyed by ocean-currents to the Arctic regions, combined with the effect of Clouds, Wind, and Aqueous Vapour, is perfectly sufficient to account for the warm and temperate condition of climate which is known to have prevailed in those regions during former epochs. The following quotations will show Sir William's views:—

"A thousand feet of depression would submerge the continents of Europe, Asia, and America, for thousands of miles from their present northern coast-lines; and would give instead of the present land-locked, and therefore ice-bound Arctic sea, an open iceless ocean, with only a number of small steep islands to obstruct the free interchange of water between the North Pole and temperate or tropical regions. That the Arctic sea would, in such circumstances, be free from ice quite up to the north pole may be, I think, securely inferred from what, in the present condition of the globe, we know of ice-bound and open seas in the northern hemisphere and of the southern ocean abounding in icebergs, but probably nowhere ice-bound up to the very coast of the circumpolar Antarctic continent, except in more or less land-locked bays. . . . . . .

"Suppose now the sea, unobstructed by land from either pole to temperate or tropical regions, to be iceless at any time, would it continue iceless during the whole of the sunless polar winter? *Yes;* we may safely answer. Supposing the depth of the sea to be not less than 50 or 100 fathoms, and judging from what we know for certain of ocean-currents, we may safely say that differences of specific gravity of the water produced by difference of temperature, not reaching anywhere down to the freezing point, would cause enough of circulation of water between the polar and temperate or tropical regions to supply all the heat radiated from the water within the Arctic circle during the sunless winter, if air contributed none of it. Just think of a current of three-quarters of a nautical mile per hour, or 70 miles per four days flowing towards the pole across the Arctic circle. The area of the Arctic circle is 700 square miles for each mile of its circumference. Hence 40 fathoms deep of such a current would carry in, per twenty-four hours, a little more than water enough to cover the whole area to a depth of 1 fathom; and this, if 7°·1 Cent. above the freezing point, would bring in just enough of heat to prevent freezing, if in twenty-four hours as much heat were radiated away as taken from a tenth of a fathom of ice-cold water would leave it ice at the freezing-point. This is no doubt much more than the actual amount of radiation, and the supposed current is probably much less than it would be if the water were ice-cold at the pole and 7°·0 Cent. at the Arctic circle. Hence, without any assistance from air, we find in the convection of heat by water alone a sufficiently powerful influence to prevent any freezing-up in polar regions at any time of year." *

That an amount of warm water flowing into the Arctic Ocean, equal to that assumed by Sir William Thomson, along with the effects of clouds, wind, dew, and other agencies to which he refers, would wholly prevent the existence of permanent ice in those regions, is a conclusion which, I think, can hardly be doubted. It is with the greatest deference that I venture to differ from so eminent a physicist; but I am unable to believe that such a transference of water from inter-tropical and temperate regions could be effected by the agency to which he attributes it. Certainly the amount of heat conveyed by means of a circulation resulting from difference of specific gravity, produced by difference of temperature, must be trifling when compared with that of ocean-currents produced by the

* Trans. of the Geol. Soc. of Glasgow, 22nd February, 1877.

impelling force of the winds. Take, for example, the case of the Gulf Stream. If the amount of heat conveyed from intertropical regions into the North Atlantic, by means of difference of density resulting from difference of temperature, were equal to that conveyed by the Gulf Stream, it would follow, as has been proved,* that the Atlantic would be far warmer in temperate and Arctic than in intertropical regions. Taking the annual quantity of heat received from the sun per unit surface at the equator as 1000, the quantities received by the three zones would be respectively as follows:—

Equator	1000
Torrid zone	975
Temperate zone . . .	757
Frigid zone	454

Assume, then, that as much heat is conveyed from intertropical regions into the Atlantic and Arctic seas by this circulation from difference of specific gravity as by the Gulf Stream, and assume also that one half of the total heat conveyed by the two systems of circulation goes to warm the Arctic Ocean, and the other half remains in temperate regions, the following would then be the relative quantities of heat possessed by the three zones:—

Atlantic	in torrid zone . . .	671
"	in temperate zone .	940
"	in frigid zone . . .	766

There is a still more formidable objection to the theory. It has been demonstrated, from the temperature-soundings made by the 'Challenger' Expedition,† that the general surface of the North

* 'Climate and Time,' Chap. XI.; "Phil. Mag.," March, 1874.

† 'Climate and Time,' pp. 220-225; "Phil Mag.," September and December, 1875; "Nature," November 25, 1875.

Atlantic must, in order to produce equilibrium, stand at a higher level than at the equator: in other words, the surface of the Atlantic is lowest at the equator, and rises with a gentle slope to well nigh the latitude of England. This curious condition of things is owing to the fact that, in consequence of the enormous quantity of warm water from intertropical regions which is being continually carried by the Gulf Stream into temperate regions, the mean temperature of the Atlantic water, considered from its surface to the bottom, is greater, and the specific gravity less, in temperate regions than at the equator. In consequence of this difference of specific gravity, the surface of the Atlantic at latitude 23°N. must stand 2 feet 3 inches above the level of the equator, and at latitude 38°N. 3 feet 3 inches above the equator. In this case it is absolutely impossible that there can be a flow in the Atlantic from the equatorial to the temperate regions resulting from difference of specific gravity. If there is any motion of the water from that cause, it must, in so far as the Atlantic is concerned, be in the opposite direction, viz., from the temperate to the equatorial regions.

All, or almost all, the heat which the Arctic seas receive from intertropical regions in the form of warm water comes from the Atlantic, and not from the Pacific; for the amount of warm water entering by Behring Strait must be comparatively small. It therefore follows from the foregoing considerations that none of that equatorial heat can be conveyed by a circulation resulting from difference of specific gravity produced by difference of temperature.

It is assumed as a condition in this theory that a submergence of the Arctic land of several hundred feet must have taken place in order to convert that land into a series of islands allowing of the free passage of water round them. But the evidence of Geology, as was shown on a former occasion, * is not altogether favourable to the idea that those warm climates were in any way the result of a submergence of the polar land. Take the Miocene epoch as an example: all the way from Ireland and the Western Isles, by the Faröes, Iceland, Franz-Joseph Land, to North Greenland, the Miocene vegetation and the denuded fragmentary state of the strata point to a much wider distribution of Polar land than that which now obtains in those regions.

Mr. Alfred R. Wallace on Mild Arctic Climates.— The theory that the mild climates of Arctic regions were due to an inflow of warm water from intertropical and temperate regions has also been fully adopted by Mr. Alfred R. Wallace. But, unlike Sir William Thomson, he does not attribute this transference of warm water to a circulation resulting from difference of density produced by difference of temperature, but to currents caused by the impelling force of the wind.

Mr. Wallace shares in the opinion, now entertained by a vast number of geologists, that during the whole of the Tertiary period the climate of the north temperate and polar regions was uniformly warm and mild, without a trace of any intervening epochs of cold. According to him, there were no glacial or interglacial periods during Tertiary times. In this case he, of course, does not suppose that the inflow of warm water into Arctic regions, on which the mild condition of climate depended, was in any way due to those physical agencies which came into operation during an interglacial period. Mr. Wallace accounts for the mild Arctic climate during the Tertiary period by the

* "Geol. Mag.," September, 1878.

supposition that at that time there were probably several channels extending from equatorial to Arctic regions through the eastern and western continents, allowing of a continuous flow of intertropical water into the Arctic Ocean. Mr. Wallace expresses his views on the point thus:—

"The distribution of the Eocene and Miocene formations shows that during a considerable portion of the Tertiary period an inland sea, more or less occupied by an archipelago of islands, extended across Central Europe, between the Baltic and the Black and Caspian Seas, and thence by narrower channels south-eastward to the valley of the Euphrates and the Persian Gulf, thus opening a communication between the North Atlantic and the Indian Ocean. From the Caspian, also, a wide arm of the sea extended during some part of the Tertiary epoch northwards to the Arctic Ocean; and there is nothing to show that this sea may not have been in existence during the whole Tertiary period. Another channel probably existed over Egypt into the eastern basin of the Mediterranean and the Black Sea; while it is probable that there was a communication between the Baltic and the White Sea, leaving Scandinavia as an extensive island. Turning to India, we find that an arm of the sea of great width and depth extended from the Bay of Bengal to the mouths of the Indus; while the enormous depression indicated by the presence of marine fossils of Eocene age, at a height of 16,500 feet in Western Thibet, renders it not improbable that a more direct channel across Afghanistan may have opened a communication between the West-Asiatic and Polar seas." *

My acquaintance with the Tertiary formations of the globe, and with the distribution of land and water

* "Island Life," p. 184.

during that period, is not such as to enable me to form any opinion whatever either as to the probability or to the improbability of the existence of such channels as are assumed by Mr. Wallace. But, looking at the question from a physical point of view, it seems to me pretty evident that if such channels as he supposes existed, allowing of a continuous flow of equatorial water into the Arctic seas, it would certainly prevent the formation of permanent ice around the Pole, and would doubtless confer on the Arctic regions a mild and equable climate. This would be more particularly the case if, as Mr. Wallace supposes, owing to geographical conditions, far more of the equatorial water was deflected into the Arctic than into the Antarctic regions.

But, at the same time, I think it is just as evident that these channels would not neutralise the effects resulting from a high state of eccentricity. It may be quite true that the physical causes, brought into operation during a high state of eccentricity, might not be sufficient to reduce the quantity of warm water flowing into the Arctic Ocean to an extent that would permit of the formation of permanent ice around the Pole, but it certainly would greatly diminish the flow into the Arctic Ocean. Supposing that, at the commencement of the last Glacial Epoch, the volume of the Gulf Stream was double what it is at present, this condition of things would not have prevented the operation of those physical agents which brought about the Glacial Epoch, although it no doubt would have considerably modified the severity of the glaciation resulting from their operation. The very same thing would hold true, though perhaps in a much greater degree, in reference to the channels assumed by Mr. Wallace.

If the emissive power of the sun was about the same during the Tertiary period as at present, and there is no good grounds for supposing it was otherwise, then the extra heat possessed by the northern temperate and Arctic regions must have been derived either from the equatorial regions or from the southern hemisphere, or, what is more likely, from both. If so, then the temperature either of the southern hemisphere or of the intertropical regions, or both, must have been much lower during the Tertiary period than at the present day. A lowering of the temperature of the equatorial regions, resulting from this transference of heat, would tend to produce a more equable and uniform condition of climate over the whole of the northern hemisphere. As the area of the Arctic Ocean is small in comparison to that of the equatorial zone, from which the warm water was derived, the fall of temperature at the equator would be much less than the rise at the pole. Supposing there had been a rise of, say, 30° at the pole, resulting from a fall of 10° at the equator (and this is by no means an improbable assumption), this would reduce the difference between the equator and the pole by 40°, or to half its present amount. We should then have a climatic condition pretty much resembling that which is known to have prevailed during at least considerable portions of the Tertiary period.

It is indeed very doubtful if such a climatic condition of things as that could be brought about by a high state of eccentricity with the present distribution of land and water; but, on the other hand, it is just as doubtful whether the channels of communication assumed by Mr. Wallace could have brought it about without the aid of eccentricity.

The very existence of so high a temperature on the

northern hemisphere during Tertiary times may be regarded as strong presumptive proof that the geographical conditions obtaining on the southern hemisphere were most unfavourable to the flow of intertropical water into that hemisphere. This may be one of the reasons why a high state of eccentricity failed to produce a well-marked glacial epoch on the northern hemisphere, the geographical conditions preventing a transference of warm water into the southern hemisphere sufficient to produce true glaciation on the opposite hemisphere. That the geographical conditions obtaining on the southern hemisphere during Tertiary times were probably of such a character is an opinion advanced by Mr. Wallace himself. "There are," he says, "many peculiarities in the distribution of plants and of some groups of animals in the southern hemisphere, which render it almost certain that there has sometimes been a greater extension of the Antarctic lands during Tertiary times; and it is therefore not improbable that a more or less glaciated condition may have been a long-persistent feature of the southern hemisphere, due to the peculiar distribution of land and sea, which favours the production of ice-fields and glaciers." *

Influences of Eccentricity during the Tertiary Period.—This being the state of things on the southern hemisphere, the glacial condition of the hemisphere, when its winter solstice was in aphelion, would tend in a powerful manner to impel the warm water of the south over on the northern hemisphere, and thus raise its temperature. This, again, is a view which has also been urged by Mr. Wallace. "Though high eccentricity would," he remarks, "not directly modify the mild climates produced by the state of the northern hemi-

* "Island Life," p. 192.

sphere which prevailed during Cretaceous, Eocene, and Miocene times,* it might indirectly affect it by increasing the mass of Antarctic ice, and thus increasing the force of the trade-winds and the resulting northward-flowing warm currents. . . . And as we have seen that during the last three million years the eccentricity has been almost always much higher than it is now, we should expect that the quantity of ice in the southern hemisphere will usually have been greater, and will thus have tended to increase the force of those oceanic currents which produce the mild climates of the northern hemisphere."†

There is little doubt but that the climate of the Tertiary period was greatly affected by eccentricity; but, owing to the difference in the geographical conditions of the two hemispheres, eccentricity would exercise a much greater influence on the climatic condition of the northern hemisphere when the northern winter solstice was in perihelion than it would do when it was in aphelion. Owing to the difference in the conditions of the two hemispheres, the physical agents brought into operation by a high state of eccentricity would act more powerfully in impelling the equatorial waters towards the Arctic regions when the winter solstice was in perihelion than they would do in impelling the water towards the Antarctic regions when the solstice was in aphelion. In this case the northern hemisphere would be heated to a greater extent when its winter solstice was in perihelion than it would be cooled when the solstice was in aphelion. It is this circum-

* High eccentricity might not directly modify the mild climates, but certainly the physical agents brought into operation by the high eccentricity would do so.

† "Island Life," p. 192.

stance which, I think, has misled geologists, and induced them to conclude that because the physical agents brought into operation when the winter solstice was in aphelion, during a high state of eccentricity, failed to produce a well-marked glacial epoch in Tertiary times, consequently the climatic condition of that period was not much affected by eccentricity.

It would seem to be owing to that peculiar difference between the conditions of the two hemispheres that, even during high eccentricity, the physical agents in operation when the winter solstice was in aphelion were unable to lower the temperature of the northern hemisphere to an extent sufficient to cover high temperate and Arctic regions with permanent ice; but for this very same reason these agents would be enabled to raise the temperature to an extent exceptionally high when the winter solstice was in perihelion. In other words, this very combination of circumstances, which so much modified the severity of what may be called the Tertiary cold periods, intensified to an exceptionally great extent the warmth and equability of what may be called the Tertiary warm periods.

CHAPTER X.

THE PHYSICAL CAUSE OF MILD POLAR CLIMATES.—*Continued.*

Climate of the Tertiary period, in so far as affected by Eccentricity.—Evidence of Alternations of Climate.—Were there Glacial Epochs during the Tertiary period?—Evidence of Glaciation during the Tertiary period.

Climate of the Tertiary period, in so far as affected by Eccentricity.—If the foregoing conclusions are correct, it is not difficult to infer what would be the probable character of the climate of the Tertiary period, in so far as that climate was affected by eccentricity. As is truly remarked by Mr. Wallace, the eccentricity during the past three million years has been almost always much higher than it is now. It will consequently follow that very considerable portions of the Tertiary age would consist of alternate comparatively cold and exceedingly warm and equable periods. These may be said to correspond to the cold and warm periods of the Glacial Epoch; but, of course, they could in no sense be called glacial and interglacial periods; for the cold of the cold periods would not be such as to produce permanent ice, while the heat and equability of the warm periods would far exceed that of the interglacial periods.

Evidence of Alternations of Climate.—That such oscillations occurred during the Tertiary period seems to be borne out by the facts of geology and palæontology. Mr. J. Starkie Gardner, a geologist who has had great experience in the fossil flora of the Tertiary deposits, says that such alternating warmer and colder conditions are supported by strong negative and some positive evidence, found not only in English Eocene, but in all Tertiary beds throughout the world. In the Lower Bagshot of Hampshire have been found, he states, feather- and fan-palms, *Dryandra*, beech, maple, *Azalea*, laurel, elm, acacia, aroids, cactus, ferns, conifers, *Stenocarpus*, and plants of the pea tribe, together with many others. The question which presents itself to one's mind, he remarks, is, how is it possible that the tropical forms, such as the palms, aroids, cactus, &c., could have grown alongside of the apparently temperate forms, such as the oak, elm, beech, and others? Mr. Gardner's explanation is as follows:—

"Astronomers, having calculated the path of the revolution of the earth in ages past, tell us that in recurring periods each hemisphere, northern and southern, has been successively subject to repeated cyclical changes in temperature. There have been for the area which is now England many alternations of long periods of heat and cold. Whenever the area became warmer, the descendants of semi-tropical forms would gradually creep further and further north, whilst the descendants of cold-loving plants would retreat from the advancing temperature, *vice versâ*. Whenever the area became gradually colder, the heat-loving plants would, from one generation to another, retreat further and further south, whilst the cold-loving plants would return to the area from which their ancestors had been driven out. In each case there would be some lingering remnants of the retreating vegetation (though perhaps existing with diminished vigour) growing alongside of the earliest arrivals of the incoming vegetation.

"Such is a possible explanation of our finding these plant-remains commingled together. It must be borne in mind that it is not so much the mean temperature of a whole year which affects the possibility of plants growing in any locality, as the fact of what are the extremes of summer and winter temperature." *

This is precisely the explanation given of the commingling of sub-tropical and Arctic floras and faunas in deposits belonging to the Glacial Epoch. The causation in the two cases was, in fact, the same in principle, differing only in the conditions under which it operated. In the case of the Glacial Epoch the cold periods were intensely severe and the warm periods but moderately hot; whereas in regard to the Tertiary cold periods they were but moderately cool, and the warm periods exceedingly hot.

Mr. Wallace, who refers to Mr. Gardner's views approvingly, says:—"In the case of marine faunas it is more difficult to judge, but the numerous changes in the fossil remains from bed to bed, only a few feet and sometimes a few inches apart, may be sometimes due to change of climate; and when it is recognised that such changes have probably occurred at all geological epochs, and their effects are systematically searched for, many peculiarities in the distribution of organisms through the different members of one deposit may be traced to this cause." †

To avoid having thus to admit the existence of alternate warmer and colder periods during Tertiary times, Mr. Searles V. Wood, Jun., proposed another theory, which is thus stated in his own words:—

"The remains upon which the determinations of this flora have been based are drifted, and not those of a bed *in situ*

* "Geological Magazine," 1877, p. 25.

† "Island Life," p. 197.

like the coal seams, and the whole of the Hampshire Eocene is connected with the delta of a great river which persisted throughout the accumulation of the various beds, which aggregate to upwards of 2000 feet in thickness. This river evidently flowed from the west, through a district of which the low ground had a tropical climate; but like some tropical rivers of the present day, such as the Brahmaputra, the Megna, the Ganges, &c., it was probably fed by tributaries flowing from a mountain region supporting zones of vegetation of all kinds from the tropical to the Arctic, if during the Eocene period vegetation such as the present Arctic had come into existence, of which we have as yet no evidence. Torrential floods may have swept the remains of vegetation from the temperate zones of this region into tributaries that conveyed it into the main river before it was decayed or water-logged, where it became intermingled with the remains of vegetation which grew in the tropical low ground skirting the main stream, so that both sank together into the same mud and silt." *

The elevated mountain regions from which he supposes these temperate forms were derived, he thinks, might have been Mull, 400 miles N.N.W., and Wales 200 miles N.W. of the Hampshire deposits. Mr. Gardner, however, showed most conclusively that Mr. Wood's theory was based on imperfect acquaintance with the conditions of the problem. The following is Mr. Gardner's reply:—

"The leaves have never been drifted from afar; they are often still adhering to the twigs. The leaves are flat and perfect, rarely even rolled and crumpled, as dry leaves may be, if falling on a muddy surface; still more rarely have they fallen edgeways and been embedded vertically. They are, moreover, not variously mixed, as they should be if they had been carried for any distance, but are found in local groups of species. For example, all the leaves of

* "Geological Magazine," 1877, p. 96.

Castanea have been found in one clay patch, with *Iriartea* and *Gleichenia;* none of these have been found elsewhere. A tri-lobed leaf is peculiar to Studland; the Alum Bay *Aralia*, the peculiar form of *Proteaceæ*, the great *Ficus*, and other leaves occur at Alum Bay only. Each little patch at Bournemouth is characterised by its own peculiar leaves. Such a distribution can only result from the proximity of the trees from which the leaves have fallen. The forms of most temperate aspect are best preserved, so that, to be logically applied, the Drift theory requires the palms, &c., to have been drifted upwards. To suppose that most delicate leaves could have been brought by torrents 400 miles from Mull or 200 miles from Wales, and spread out horizontally in thousands, without crease or crumple, on the coast of Hampshire, may be a feasible theory to Mr. Searles V. Wood, Jun., but will not recommend itself to the majority of thinkers." *

Were there Glacial Epochs during the Tertiary Age?—Many geologists, especially amongst those who are opposed to the theory of recurring glacial epochs, answer this question emphatically in the negative. This belief as to the non-existence of glacial conditions during the Tertiary period is, of course, based wholly on negative evidence; and this negative evidence, though strong, is by no means perfectly conclusive, and certainly not worthy of the weight which has been placed upon it. In Chapter XVII. of 'Climate and Time' I have endeavoured to show that, although much has been written on the imperfection of geological records, yet the imperfection of those records in regard to past glacial epochs has not received the attention which it really deserves.

It must be borne in mind, however, that it does not follow from the physical theory of secular changes of

* "Geological Magazine," 1877, p. 138.

climate, that because the eccentricity may have been high at some particular period there must necessarily have been a glacial epoch. The erroneous nature of this misapprehension of the theory has already been shown at considerable length in Chapter V. Eccentricity can produce glaciation only through means of physical agencies, and for the operation of these agencies a certain geographical condition of things is absolutely necessary. We know with certainty that during the Tertiary period the eccentricity was at times exceptionally high, as, for example, 2,500,000 and 850,000 years ago; but whether a glacial epoch occurred at these periods depended, of course, upon whether or not the necessary geographical conditions then obtained. Supposing the necessary geographical conditions for glaciation did exist at the two periods in question, still if these conditions differed very much from those which now obtain, the glacial state of things then produced would certainly differ from that of the last glacial epoch. This is obvious, for the same physical agencies acting under very different conditions would not produce the same effects. Under almost any geographical condition of things eccentricity would produce marked effects, but the effects produced might not amount to glaciation. In the Tertiary age, during high eccentricity, the effects resulting might possibly have been as well marked as they were during the Glacial Epoch; but these effects must have differed very much from those produced at that epoch. We have seen that, owing to that peculiar geographical condition of things existing during the Tertiary period, the physical agents brought into operation by a high state of eccentricity would have a much greater influence in *raising* the temperature of the northern hemisphere when the winters occurred in perihelion, than they would have in *lowering* the temperature of that hemisphere when the winters were in aphelion. At the periods 2,500,000 and 850,000 years ago, for example, those physical agents would no doubt produce quite a tropical condition of climate in high northern latitudes when the winters were in perihelion, while it is quite probable they may not have been able to produce glaciation when the winters were in aphelion. It is more than likely that the tropical nature of the climate during portions of the Tertiary period was due not so much to that peculiar distribution of land and water existing then, as it was to the fact that this peculiar distribution enabled the physical agents in operation during a high state of eccentricity to impel a vastly greater amount of warm intertropical water into the temperate and Arctic seas than they could have done under the present geographical condition of things.

Those very same geographical conditions enabling the physical agents to raise the temperature exceptionally high during the warm periods would, on the other hand, prevent them from being able to lower the temperature exceptionally low during the alternate cold periods. Nevertheless, I think it probable that at the two periods referred to, when the eccentricity was much greater than it was during the Glacial Epoch, the temperature would be lowered to an extent that would produce glaciation, although the glaciation might not perhaps approach in severity to anything like that of the Glacial Epoch. The negative evidence which has been adduced against the existence of such glacial conditions during the Tertiary period is certainly far from being conclusive.

The opinion is concurred in by Mr. Wallace that the Table of Eccentricity for the past three million years, given in 'Climate and Time,' probably includes the greater part, if not the whole, of the Tertiary period. He states that during the 2,400,000 years preceding the last Glacial Epoch there were, according to the table, no fewer than sixteen separate epochs when the eccentricity reached or exceeded twice its present amount. But it does not follow, according to the physical theory, that there ought, on that account, to have been sixteen epochs of more or less glaciation. Whether such ought to have been the case or not would depend on whether or not the geographical conditions were the same during these epochs as they were at the Glacial Epoch, a consideration with which the theory has no relation. The question is not, were there sixteen glacial epochs during the Tertiary period, but were there any? Even granted that those channels assumed by Mr. Wallace did exist, they would not, I feel assured, wholly prevent glaciation taking place at the two periods to which reference has been made, although the glaciation might not be severe.

In support of the opinion that there is no evidence of glaciation during the Tertiary period, Mr. Wallace quotes the views of Mr. Searles V. Wood, Jun., on the subject. Mr. Wood states that the Eocene formation is complete in England, and is exposed in continuous section along the north coast of the Isle of Wight and along the northern coast of Kent from its base to the Lower Bagshot Sand. It has, he says, been intersected by cuttings in all directions and at all horizons, but has not yielded a trace of anything indicating a cold and glacial condition of things. The same, he adds, holds true of the strata in France and Belgium. Further, "the Oligocene of Northern Germany and Belgium, and the Miocene of those countries and of France, have also afforded a rich molluscan fauna, which, like that of the Eocene, has as yet presented no indication of the intrusion of anything to interfere with its uniformly sub-tropical character."

In reply to all this, it may be stated that the simple absence of any trace of glaciation in the Tertiary deposits of the South of England certainly cannot be regarded as conclusive against the existence of an epoch of glaciation during that period. Not many years ago geologists denied that there was any evidence to be found of glaciation in the South of England, and at the present time there are hundreds of geologists who will not admit that that part was ever overridden by land-ice. If it is so difficult to find in that quarter evidence of the last glacial epoch, severe as that glacial epoch was, we need not wonder that no trace of glaciation so remote as that of the Eocene period is now to be seen. Besides all this, there is in the South of England the land-surface on which the glaciation, if any, took place, whereas not a vestige of the old land-surfaces of the Eocene period now remains. It therefore seems to me that the mere fact of nothing as yet having been found in the Tertiary deposits of the South of England, indicating one or more cold periods, is no proof that there may not possibly have been such periods, and even of considerable severity. The same remarks hold equally true in regard to the deposits on the continent referred to by Mr. Wood.

It will be urged in reply that there is one kind of evidence which ought to be found, as it could not possibly have been obliterated by the destruction of the Tertiary land-surfaces—that is, the presence of erratic blocks and foreign rock-fragments imbedded in the strata. Mr. Wallace states that in the many thousand feet in thickness of alternate clays, sands, marls, shales, and limestones, no irregular blocks of foreign material or boulders characteristic of glacial conditions are to be found. The same, he says, holds equally true of the extensive Tertiary deposits of temperate North America.

If it be really the case that the Tertiary beds are wholly without boulders or fragments of foreign material, then this certainly may be regarded as proof that no real glacial epoch could have occurred during that period. But has it been satisfactorily ascertained that those beds are wholly devoid of such materials? Those beds, I presume, have been searched by geologists for their fossil contents rather than for stratigraphical evidence of glacial epochs. It is remarkable how long the evidence of glaciation sometimes remains unobserved when no special attention is devoted to the matter. As examples of this, we know with certainty that the Orkney and Shetland Islands were during the Glacial Epoch overridden by land-ice; and yet geologists who had often visited these islands declared that they bore no marks of glaciation. So recently as 1875 the low grounds of Northern Germany were believed to be without glacial striæ; yet when German geologists began to turn their attention specially to the subject, they found not only evidence of glaciation but indisputable proof that during the Glacial Epoch the great Scandinavian ice-sheet had advanced over the country no fewer than three separate times down to the latitude of Berlin. I have myself seen the striated summit of a mountain on which geologists had been treading for years without observing the ice-markings under their feet. The reason why these markings so long escaped detection is doubtless due to the fact that they were on a spot which no geologist supposed that land-ice could have reached. For this very same reason the fact remained so long unobserved that the low-lying ground of Caithness had been glaciated by land-ice from Scandinavia, filling the entire Baltic and the North Sea. Many similar cases might be adduced where the marks of glaciation remained long unobserved, either because no special search had been made for them, or because they were under conditions in which they were not expected to be found. It is very probable that when the Tertiary deposits are carefully examined, with the special object of ascertaining whether or not they contain evidence of glaciation, geologists may be led to a different conclusion regarding the supposed uniformly warm character of the climate of that period. They may possibly find that, after all, the Tertiary beds do contain boulders and foreign material, indicating the existence of glacial conditions during the period.

Considerable importance has been attached to the statement of Professor Nordenskjöld that he failed to observe in the stratified deposits of Greenland and Spitzbergen any evidence whatever of former glaciation in those regions. "We have never seen," he says, "in Spitzbergen nor in Greenland, in these sections often many miles in length, and including, one may say, all formations from the Silurian to the Tertiary, any boulders even as large as a child's head. There is not the smallest probability that strata of any considerable extent containing boulders are to be found in the polar tracts previous to the middle of the Tertiary period. Both an examination of the geognostic condition and an investigation of the fossil flora and fauna of the polar lands, show no signs of a glacial era having existed in those parts before the termination of the Miocene period."* That Professor Nordenskjöld

* "Geological Magazine," 1875, p. 531.

may not have seen in those strata boulders larger than a child's head may be perfectly true, but that there actually are none is a thing utterly incredible. Still more incredible, however, is the conclusion which he draws from this absence of boulders—viz., that from the Silurian down to the termination of the Miocene period no glacial condition of things existed either in Greenland or in Spitzbergen. Both these places are at present in a state of glaciation; and were it not, as we have seen, for the enormous quantity of heat which is being transferred from the equatorial regions by the Gulf Stream, not only Greenland and Spitzbergen, but the whole of the Arctic regions would be far more under ice than they are. A glacial state of things is the normal condition of polar regions; and if at any time, as during the Tertiary age, the Arctic regions were free from snow and ice, it could only be in consequence of some peculiar distribution of land and water and other exceptional conditions. That this peculiar combination of circumstances should have existed during the whole of that immense lapse of time between the Silurian and the close of the Tertiary period is certainly improbable in the highest degree. In short, that Greenland during the whole of that time should have been free from snow and ice is as improbable, although perhaps not so physically impossible, as that the interior of that continent should at the present day be free from ice and covered with luxuriant vegetation. Perhaps the same skill and indomitable perseverance which proved the one conclusion to be erroneous may yet one day prove the other to be also equally mistaken.

Professor Nordenskjöld does not appear to believe in alternations of climate even in temperate regions for he says, "from palæontological science no support, can be obtained for the assumption of a periodical alternation of warm and cold climates on the surface of the earth."

Evidence of Glaciation during the Tertiary Period. —Evidence of glaciation during the Miocene period is, I think, afforded by the well-known conglomerates and erratics near Turin, first described by M. Gastaldi. Beds of Miocene sandstone and conglomerate, with an intercalated deposit containing large angular blocks of greenstone and limestone have been found. Some of these blocks are of immense size. Many of the stones in the deposit are polished and striated in a manner similar to those found in the boulder-clay of this country. It has been shown by Gastaldi that these blocks have all been derived from the outer ridge of the Alps on the Italian side—namely, from the range extending from Ivrea to the Lago Maggiore, and, consequently, they must have travelled from twenty to eighty miles. So abundant are these large blocks that extensive quarries have been opened in the hills for the sake of procuring them. The stratification of the beds amongst which the blocks occur sufficiently indicates aqueous action and the former presence of the sea. That glaciers from the southern Alps actually reached to the sea and sent adrift their icebergs over what are now the sunny plains of Northern Italy, is proof that during that cold period the climate must have been very severe. One remarkable circumstance, indicating not only the glacial condition of the bed in which the blocks occur, but also that this glaciation was the result of eccentricity, is the fact that the bed is wholly destitute of organic remains, while they are found abundantly both in the underlying and overlying beds.

Evidence of glaciation during the Eocene period, as

is also well known, is found in the "*Flysch*" of Switzerland. On the north side of the Alps, from Switzerland to Vienna, and also near Genoa, there is a sandstone a few thousand feet in thickness, containing enormous blocks of Oolitic limestone and granite. Many of these blocks are upwards of 10 feet in length, and one at Halekeren, near the lake of Thun, is 105 feet long, 90 feet broad, and 45 feet in thickness. The block is of a granite of a peculiar kind, which cannot be matched anywhere in the Alps. Similar blocks are found in beds of the same age in the Appennines and in the Carpathians. The glacial origin of this deposit is further evinced by the fact that it is wholly destitute of organic remains. One circumstance, which indicates that this glaciation was due to eccentricity, is the fact that the strata most nearly associated with the "*Flysch*" are rich in Echinoderms of the *Spatangus* family, which have a decided tropical aspect. This is what we ought, of course, *à priori*, to expect if the glaciation was the result of eccentricity, for the more severe a cold period of a glacial epoch is, the warmer will be the periods which immediately precede and succeed.

Some writers endeavour to account for those glacial phenomena, without any reference to the influence of high eccentricity, by the assumption that the Alps were much more elevated during the Tertiary period than they are at the present day. If we, however, adopt this explanation, we shall have to assume that the Alps were suddenly elevated at the time when the bed containing the erratics began to be deposited, and that they were as suddenly lowered when the deposition of the bed came to a close—a conclusion certainly very improbable. Had the lowering of the Alps been effected by the slow process of denudation, it must have taken a long course of ages to have lowered them to the extent of bringing the glacial state to a close. In this case there ought to be a succession of beds indicating the long continuance of cold conditions. Instead of this, however, we have a glacial bed immediately preceded and succeeded by beds indicating an almost tropical condition of climate. When we take this circumstance into consideration, along with the evidence adduced by Mr. J. S. Gardner as to the alternations of warmer and colder conditions in the south of England and other parts of Europe during the Eocene period, the conviction is forced upon us that a high state of eccentricity is the most rational explanation of these curious phenomena.

The greater elevation of the Alps would undoubtedly intensify the glacial condition of things, but it would not originate it. The elevated character of the Alps, for example, was no doubt the reason why the plains of Switzerland, during the last glacial epoch, were so much more buried under ice than other parts of Southern Europe; but their elevation was not that which brought about the glaciation, for those plains were free from ice both before and after the glacial epoch, though the Alps were no doubt as high as they were during the ice-period.

If we adopt the theory that these glacial conditions were due to eccentricity, then we have, as I endeavoured to show many years ago,* a clue to the probable absolute date of the Middle Eocene and the Upper Miocene periods. There were, as we have seen, two epochs during the Tertiary period when the eccentricity was exceptionally high, viz., 2,500,000 years ago and 850,000 years ago. The former might probably be the

* "Phil. Mag.," November, 1868; 'Climate and Time,' chap. xxi.

date of the "*Flysch*" of the Eocene formation, and the latter the date of the period when the Miocene erratics were deposited in the icy sea near Turin.

Some geologists have maintained that the climatic conditions of the Tertiary period are utterly hostile to the Physical Theory of Secular changes of Climate. The very reverse, however, is the case; for, as we have seen, several of the facts of Tertiary climate can be explained on no other principle than that of the theory.

CHAPTER XI.

INTERGLACIAL PERIODS IN ARCTIC REGIONS.

Interglacial Periods in Arctic Regions more marked than Glacial.—Evidence from the Mammoth in Siberia.—Northern Siberia much Warmer during the Mammoth Epoch than now.—Evidence from Wood.—Evidence from Shells.—The Mammoth Interglacial.—Main Characteristics of Interglacial Climate.—Evidence from the Mammoth in Europe.—The Mammoth Glacial as well as Interglacial.—Arctic America during Interglacial Times.—Was Greenland Free from Ice during any of the Interglacial Periods?

IN Chapter VIII. of the present volume, and also in 'Climate and Time' (Chapter XVI.), it was pointed out that in temperate regions the cold periods of the Glacial Epoch would be far more marked than the warm interglacial periods. In temperate regions the condition of things which prevailed during the cold periods would differ far more widely from that which now prevails than would the condition of things during the warm periods. But as regards the polar regions the reverse would be the case; there the warm interglacial periods would be more marked than the cold periods. The condition of things prevailing in these regions during the warm periods would be in strongest contrast to what now obtains; but this would not hold true in reference to the cold periods, during which matters would be pretty much the same as at present, only somewhat more severe. In short, the glacial state is the normal condition of the polar regions, the interglacial the abnormal. At present Greenland and other parts of the Arctic regions are almost wholly covered with snow and ice, and, consequently, nearly destitute of vegetable life. In fact, as regards organic life in those regions, matters during the Glacial Epoch would not probably be much worse than they are at the present day. Greenland and the Antarctic continent are to-day almost as destitute of plant life as they could possibly be. Although, in opposition to what is found to be true in reference to the temperate regions, the polar interglacial periods were more marked than the glacial, it does not follow that on this account the relics of the interglacial periods which remain ought to be more abundant in polar than in temperate regions. On the contrary, the reverse ought to be the case. In the polar regions, undoubtedly, there is least likelihood of finding traces of interglacial periods; for there, of all other places, the destruction of such traces would be most complete. The more severe the glaciation following a warm period, the more complete would be the removal of the remains belonging to the period. If, in such places as Scotland and Scandinavia, so little is left of the wreck of interglacial periods, it need be a matter of no surprise that in Arctic regions scarcely a relic of those periods remains. The comparative absence in polar regions of organic remains belonging to a mild interglacial period cannot therefore be adduced as evidence against the probable existence of such a period. Who would expect to find such remains in ice-covered regions like Greenland and Spitzbergen? Although not a trace is now to be found, it is nevertheless quite possible that during interglacial periods those regions may have enjoyed a comparatively mild and equable climate.

Evidence from the Mammoth in Siberia.—This comparative absence of the remains of a warmer condition of climate in Arctic regions during Pleistocene times holds true, however, only in regard to those parts, like Greenland, which have undergone severe glaciation. When we examine Siberia and other places which appear to have escaped the destructive power of the ice, we find, from a class of facts the physical importance of which appears to have been greatly overlooked, abundant proofs of a mild and equable condition of climate. I refer to facts connected with the climatic condition under which the Siberian mammoth and his congeners lived. The simple fact that the mammoth lived in Northern Siberia proves that at the time the climate of that region must have been far different from what it is at the present day.

The opinion was long held, and is still held by some, that the mammoth did not live in Northern Siberia, where his remains are found, but in more southern latitudes, and that these remains were carried down by rivers. It was considered incredible that an animal allied to the elephant, which now lives only in tropical regions, should have existed under a climate so rigorous as that of Siberia. The opinion that the remains were floated down the Siberian rivers is now, however, abandoned by Russian naturalists and other observers who have carefully examined the country.

I shall here give a brief statement of the facts and arguments which have been adduced in support of the theory that the mammoth lived and died where its remains were found. For these facts I am mainly indebted to the admirable papers by Mr. Howorth on the "Mammoth in Siberia," which appeared in the "Geological Magazine" for 1880.

Had the remains of the mammoth been carried down from the far south by the Siberian rivers, they would have been found mainly, if not exclusively, on the banks of the long rivers, such as the Obi, Yenissei, and the Lena, and in the deltas formed at their mouths. But such is not the case. "These are," says Mr. Howorth, "found even more abundantly on the banks of the very short rivers east of the Lena. They are found not only on the deltas of these rivers, but far away to the north, in the islands of New Siberia, beyond the reach of the currents of the small rivers, whose mouths are opposite those islands. But a more convincing proof is that "they are found not only in North Central Siberia, where the main arteries of the country flow, but in great numbers east of the river Lena, and in the vast peninsula of the Chukchi, in the country of the Yukagirs, and in Kamtskatka, where there are no rivers down which they could have floated from more temperate regions." Besides, it is not merely in the deltas and banks of rivers that the remains are found, but in nearly all parts of the open tundra; and Wrangell says* that the best as well as the greatest number of remains are found at a certain depth below the surface in clay-hills, and more in those of some elevation than along the low coast or in the flat tundra.

Had the mammoth lived in the south we should, as Mr. Howorth further remarks, have found its remains most abundant in the south, whereas the farther north we go the remains become more abundant; and in the islands of the Liachof archipelago, in about latitude 74°, the greatest quantities have been discovered. Again, according to Hedenstrom, the bones and tusks found in the north are not so large and heavy as those

* "Polar Sea Expedition," English translation, p. 275.

in the south—a fact which still further confirms the opinion that the mammoth lived where his remains are found, inasmuch as the greater severity of the climate in northern parts would certainly hinder the growth and full development of the animal.

Northern Siberia much Warmer during the Mammoth Epoch than now.—It is true that the mammoth and the *Rhinoceros tichorhinus* were furnished with a woolly covering which would protect them from cold; but it is nevertheless highly improbable that they could have endured a climate so severe as that of Northern Siberia at the present day, where the ground is covered with snow for nine months in the year, and the temperature is seldom much above zero Fahr. And even if they could have endured the cold, they would have starved for want of food. Some parts of Siberia are no doubt fertile, as, for example, the valley of the Yenissei, described by Nordenskjöld;* but there is little doubt, as Mr. Howorth remarks, that the larger portion of Northern Siberia, where the mammoth and the rhinoceros lived, is now a naked tundra covered with moss, on which no tree will grow. On such ground it is physically impossible that the mammoth and rhinoceros could exist, for they cannot graze close to ground like oxen. They live on long grass, and on the foliage and small branches of trees.

Evidence from Wood.—The fact that the mammoth was most abundant beyond the present northern limit of wood is pretty good evidence that the climatic condition of Northern Siberia must then have been milder than now. Wood must have extended, in the days of the mammoth, far beyond its present limit, probably as far north as New Siberia: facts of observation support this conclusion.

* "Nature," Dec. 2, 1875.

The wood found in Northern Siberia consists of two classes—the one is the result of drift, the other grew on the spot. The natives call the former "Noashina," and the latter "Adamshina;" and the division is supported by Göppert, who separates the trunks of timber found in Northern Siberia into a northern series, with narrow rings of annual growth, and a southern, with wider ones. The latter doubtless floated down the rivers, as great quantities do still; while the former probably grew there with the mammoth.

In the middle of October, 1810, Hedenstrom went across the Tundra direct to Ustiansk. "On this occasion," he says, "I observed a remarkable natural phenomenon on the Chastach Lake. This lake is 14 versts long and 6 broad, and every autumn throws up a quantity of bituminous fragments of wood, with which its shores in many places are covered to the depth of more than 2 feet. Among these are pieces of a hard transparent resinous substance, burning like amber, though without its agreeable perfume. It is probably the hardened resin of the larch tree. The Chastach Lake is situated 115 versts from the sea and 80 versts from the nearest forest." *

On the same journey, Hedenstrom noticed "on the Tundra, equally remote from the present line of forest, among the steep sandy banks of the lakes and rivers, large birch trees, complete, with bark, branches, and root. At the first glance, they appeared to have been well preserved by the earth, but on digging them up, they are found to be in a thorough state of decay. On being lighted they glow, but never emit a flame; nevertheless the inhabitants of the neighbourhood use them as fuel, and designate these subterranean trees

* Wrangell, "Polar Sea Expedition," p. 491.

as *Adamoushtshina*, or of Adam's time. The first living birch tree is not found nearer than three degrees to the south, and then only in the form of a shrub." *

On the hills in the interior of the island of Koteloni, "Sannikow found the skulls and bones of horses, buffaloes, oxen, and sheep in such abundance that these animals must formerly have lived there in large herds. At present, however, the icy wilderness produces nothing that could afford them nourishment, nor would they be able to endure the climate. Sannikow concludes that a milder climate must formerly have prevailed here, and that these animals may therefore have been contemporary with the mammoth, whose remains are found in every part of the island." †

"Herr von Ruprecht reported to Brandt that, at the mouth of the Indiga, in 67° 39′ N. lat., on a small peninsula called Chernoi Noss, where at present only very small birch bushes grow, he found rotten birch trunks still standing upright, of the thickness of a man's leg and the height of a man. In going up the river, he met with no traces of wood until he reached the port of Indiga. Here he noticed the first light-fir woods growing among still standing but dead trunks. And higher up the river still, the living woods fairly began." ‡

Schmidt says that, "where the lakes on the Tundra have grown small and shallow, we find on and near their banks a layer of turf, under which, in many places, *are remains of trees in good condition, which support the other proofs that the northern limit of trees has retrogressed, and that the climate here has grown colder.* I found, on the way from Dudino to

* Wrangell, "Polar Sea Expedition," p. 492.

† Ibid, p. 496.

‡ Bull. of Soc. of Nat. of Moscow; quoted by Howorth.

the Ural Mountains, in a place where larches now only grow in sheltered river-valleys, in turf on the top of the tundra, prostrate larch trees still bearing cones." *

Schmidt also states that he was informed that at Dudino, just at the limit of the woods, there had been found in a miserable larch wood the lower part of a stem sticking in the ground, apparently rooted, which was three feet in diameter. He also states that, "eleven versts above Krestowkoje, in lat. 72°, he found, in a layer of soil covered with clay, on the upper edge of the banks of the Yenissei, well-preserved stems like those of the birch with their bark intact, and sometimes with their roots attached, and three to four inches in diameter. Professor Merklin recognises them as those of the *Alnaster fruticosus*, which still grows as *a bush* on the islands of the Yenissei, in lat $70\frac{1}{2}$° N."

Evidence from Shells.—In the fresh-water deposits in which the bones of the mammoth are found, there are fresh-water and land-shells, which indicate a warmer condition of climate. I quote the following from Mr. Howorth's memoir:—

"Schmidt found *Helix schrencki* in fresh-water deposits on the Tundra below Dudino and beyond the present range of trees. Lopatim found recent shells of it, with well-preserved colours, 9° farther south, in lat. 68° and 69°, within the present range of trees, at the mouth of the Awamka. The most northern limit hitherto known for this shell was in lat. 60° N., where they were found by Maak in gold-washings on the Pit."

"In the fresh-water clay of the Tundra, by Tolstoi Noss, Schmidt found *Planorbis albus*, *Valvata cristata*, and *Limnæa auricularia* in a sub-fossil state; *Cyclas calyculata* and *Valvata piscinalis* he found thrown up

* Schmidt, as quoted by Howorth.

on the banks of the Yenissei, and on a rotten drifted trunk *Limax agrestis; Anodonta anatina* he also found on the banks of the Yenissei as far as Tolstoi Noss, but no farther. *Pisidium fontinale* still lives in the pools on the Tundra; as does *Succinea putris* on the branches of the Alnaster on the Brijochof Islands."

Mr. Belt mentions* that the *Cyrena fluminalis* is found in Siberia in the same deposits which contain the remains of the mammoth and the *Rhinoceros tichorhinus.*

"The evidence, then," says Mr. Howorth, "of the *debris* of vegetation, and of the fresh-water and land-shells found with the mammoth remains, amply confirms the *à priori* conclusion that the climate of Northern Siberia was at the epoch of the mammoth much more temperate than now. It seems that the botanical facies of the district was not unlike that of Southern Siberia; that the larch, the willow, and the Alnaster were probably the prevailing trees, that the limit of woods extended far to the north of its present range and doubtless as far as the Arctic Sea; that not only *the mean temperature was much higher, but it is probable that the winters were of a temperate and not of an Arctic type.*" †

The Mammoth Interglacial.—It need be a matter of no surprise that the climate of Northern Siberia during the time of the mammoth was more mild and equable than now, if we only admit that the mammoth was interglacial. That it was of interglacial age is a conclusion which, I think, has been well established by Professor J. Geikie and others. Into the facts and arguments which have been advanced in support of this conclusion I need not here enter. The subject

* "Quart. Journ. Geol. Soc.," Vol. xxx., p. 464.

† "Geol. Mag." December 1880.

will, however, be found discussed at great length in Professor Geikie's "Prehistoric Europe" and in "The Great Ice Age" (second edition). Mr. R. A. Wallace considers that one of the last intercalated mild periods of the Glacial Epoch seems to offer all the necessary conditions for the existence of the mammoth in Siberia. That the mammoth was interglacial will be further evident when we consider the climatic conditions of Europe at the time that it lived there. Before doing so, it may be as well to glance at what evidently were the main characteristics of the interglacial periods.

Main Characteristics of Interglacial Climate.—They are as follows:—

1. Interglacial conditions neither did nor could exist *simultaneously* on both hemispheres. They existed only on one hemisphere at a time, viz., on the hemisphere which had its winter solstice in perihelion.

2. During interglacial periods the climate was more *equable* than it is at present; that is to say, the difference between the summer and winter temperatures was much less than it is now. The summers may not have been warmer or even so warm as they are at present, but the temperature of the winters was much above what it is at the present day.

3. During interglacial periods the quantity of equatorial heat conveyed by ocean-currents into temperate and polar regions was far in excess of what it is at present. On this account a greater *uniformity* of climate then prevailed; that is to say, the difference of climatic conditions between the sub-tropical and the temperate and polar regions was less marked than at present—the temperature not differing so much with latitude as it now does.

4. *Mildness*, or a comparative absence of high winds, characterised interglacial climate. This partial exemp-

tion from high winds resulted from the fact that the difference of temperature between the equator and the poles, the primary cause of the winds, was much less than at the present day.

5. Another character of interglacial climate was a *higher mean temperature* than now prevails. This, amongst other causes, resulted from the great amount of heat then transferred by ocean-currents from the glacial to the interglacial hemisphere.

6. During interglacial periods the climate was not only more equable, mild, and uniform than now, but it was also more *moist*. This was, doubtless, owing mainly to the fact of the presence then in temperate and polar regions of so large an amount of warm intertropical water. In short, it was the presence of so much warm water from intertropical regions which mainly gave to the climate of the interglacial periods its peculiar character.

All these characteristics of interglacial climate have been fully established by the facts of geology, but they are also, as we have seen, deducible *à priori* from physical principles. They follow as *necessary consequences* from those physical agencies which brought about the glacial epoch.

Evidence from the Mammoth in Europe.—Skeletons and detached remains of the mammoth have been found in nearly every country in Europe. Mr. Howorth, in his memoir,* gives the details of the finding of these in various parts of Russia, Germany, Denmark, Sweden, Belgium, France, England, and other countries. It is shown that the conditions under which the mammoth remains have been found in Europe are almost identically the same as those under which they are found in Siberia, with the exception, of course,

* "Geol. Mag.," May, 1881.

that in Europe no carcases with their flesh intact have been met with.

Again, the deposit in which the mammoth remains are found in Europe is the same as that in which they occur in Siberia. The deposit is a fresh-water one, consisting of marly clay and gravel, and containing plant remains and land and fresh-water shells. When these plants and shells are examined, they are found to indicate the same interglacial condition of climate as that which prevailed in Siberia during the time the mammoth lived in that region.

In the case of land-plants it is, of course, only under exceptional circumstances, as Professor J. Geikie remarks, that they can be found in a condition suitable for the botanist. Now and again, however, beds with well-preserved plants are met with, buried under lacustrine deposits. In a still better state of preservation are the plant-remains and shells which have been discovered in the masses of calcareous tufa which have been formed upon the borders of incrustating springs. An examination of the plant-remains found under those conditions shows that during Pleistocene times, when the deposits in which the mammoth bones are found were being formed, the climate was more equable and uniform than it is at the present day.

The fossiliferous remains yielded by the tufas have led to most important results as to the climatic condition of the Pleistocene period, into the details of which I need not here enter. These will be found at full length in Professor J. Geikie's "Prehistoric Europe," Chapter IV.* It will suffice at present simply to refer to the general conclusions to which these researches have led, in so far as they bear on the climatic conditions prevailing at the time the mammoth lived so abundantly in Europe.

* See also Mr. Howorth's memoir, "Geol. Mag.," June, 1881.

In the tufa deposits of Tuscany have been found numbers of plant-remains of indigenous species, commingled with others which now no longer grow in Tuscany. Amongst the latter is the Canary laurel, which now flourishes so luxuriantly in the Canary Islands, on the northern slopes of the mountains, at an elevation of from 2000 to 5000 feet above the sea-level —a region, remarks Professor J. Geikie, nearly always enveloped in steaming vapours, and exposed to heavy rains in winter. In that deposit is also found the common laurel, associated with the beech. This is not now the case, as the laurel requires more shade than it can find there at present, while the beech has retreated to the northern flanks of the Appennines to obtain a cooler climate.

In the tufas of Provence are found groups the same as those which flourish there at present, but commingled with them are also the Canary laurel and other plants which are no longer natives of Provence. Saporta directs attention to the fact that species such as the Aleppo pine and the olive, demanding considerable summer-heat rather than a moist climate, are entirely wanting in the tufas.

Similar to those of Provence are the tufas of Montpellier. Saporta concludes that when all those species lived together the climate must necessarily have been *more equable and humid* than at present. In other words, the summers were not so dry and the winters were milder than they are now.

The deposit near Moret, in the valley of the Seine, is still more remarkable in showing the equable condition of climate which then prevailed. The assemblage of plants found there tells a tale, says Professor J. Geikie, which there is no possibility of misreading. "Here," he says, "we have the clearest evidence of a genial, humid, and equable climate having formerly characterised Northern France. The presence of the laurel, and that variety of it which is most susceptible to cold, shows us that the winters must have been mild, for this plant flowers during that season, and repeated frosts, says Saporta, would prevent it reproducing its kind. It is a mild winter rather than a hot summer which the laurel demands, and the same may be said of the fig-tree. The olive, on the other hand, requires prolonged summer heat to enable it to perform its vital functions. Saporta describes the fig-tree of the La Celle tufa as closely approximating, in the size and shape of its leaves and fruit, to that of the tufas in the south of France, and to those of Asia Minor, Kurdistan, and Armenia. But if the winters in Northern France were formerly mild and genial, the summers were certainly more humid, and probably not so hot. This is proved by the presence of several plants in the tufa of La Celle which cannot endure a hot arid climate, but abound in the shady woods of Northern France and Germany."

The plants found in the tufas of Canstadt are much similar to those of Moret. Mr. Howorth, in regard to the deposits of those places, says:—"The co-existence of the species found there, remarks M. Saporta, proves very clearly that, notwithstanding the variations due to latitude, Europe, from the Mediterranean to its central districts, offered fewer contrasts, and was more uniform than it is now. A more equable climate, damp and clement, allowed the *Acer pseudo-platanus* and the fig to live associated together near Paris, as it allowed the reindeer and hyæna. The *Acer* grows with difficulty now where the *Ficus* grows wild, while the latter has to be protected in winter in the latitude of Paris."*

* "Geol. Mag." June, 1881.

Equally conclusive is the testimony borne by the Mollusca of the tufas. In the tufas and marls of Moret, in the valley of the Seine, 35 species were discovered. The majority of these must have lived in damp and shady places, in the recesses of moist woods, and on the leaves of marsh-plants. The shells, M. Tournouër concludes, bespeak a condition of climate more uniform, damp, and equable than now prevails in that region, with a somewhat higher mean annual temperature. In the alluvial deposits of Canstadt, in Würtemberg, a class of shells indicating a similar condition of climate has been discovered.

The evidence furnished by the animals found most abundantly with the mammoth in Europe and Siberia, Mr. Howorth thinks, points to the same conclusion as the testimony of the plants and mollusca associated with that huge mammal.

The same mild and equable condition which allowed of the mammoth living in Northern Siberia during Pleistocene times thus equally prevailed over the whole of Europe. We have seen that, according to the Physical Theory, this condition of climate was in every respect precisely what it ought to have been on the supposition that it was interglacial. It was a condition mild, equable, uniform, humid, and of a higher mean annual temperature than we have at the present day. There is, however, direct and positive evidence that this condition of climate was interglacial; for the facts both of geology and of palæontology show that it was preceded and succeeded by a state of things of a wholly opposite character.

The Mammoth Glacial as well as Interglacial.—Although the mammoth could have lived in Arctic Siberia only during an interglacial period, it does not follow that it must have perished during the succeeding glacial period. When the cold came on, and the vegetation on which it subsisted began to disappear, it would move southwards, and would continue its march as the cold and severity of the winters increased. During the continuance of the ten or twelve thousand years of Arctic conditions it would find in Southern Europe and elsewhere places where it could exist. At the end of the cold period, and when the climate again began to grow mild and equable, it would retrace its steps northwards. There is, however, little doubt that during the severity of a glacial period, and when necessarily confined to a more limited area, its numbers would be greatly diminished. There is every reason for believing that the mammoth outlived all that succession of cold and warm periods known as the Glacial Epoch proper, and did not finally disappear till recent postglacial times.

It was probably about the commencement of a cold period, and before the mammoth had retreated from Northern Siberia, that those individuals perished whose carcases have been found frozen in the cliffs. The way in which they probably perished and became imbedded in the frozen mud and ice has, I think, been ingeniously shown by Dr. Rae. His views on the subject are as follows:—

"The mammoths drowned would float down the rivers, and would probably get aground in three or four feet of water. As soon as the winter set in, they would be frozen up in this position. The ice in so high a latitude as 70° or 75° north would acquire a thickness of five or six feet at least, so that it would freeze to the bottom on the shallows where the mammoths were anchored. In the spring, on the breaking-up of the ice, this ice being solidly frozen to the muddy bottom would not rise to the surface, but remain fixed, with its contained animal remains, and the flooded stream would rush over both, leaving a covering of mud as the water subsided. Part of this fixed ice, but not the whole, might be thawed away during summer, and next winter a fresh layer of ice with a fresh supply of animal remains might be formed over the former stratum; and so the peculiar position and perfect state of preservation of this immense collection of extinct animals may be accounted for without having recourse to the somewhat improbable theory that a very great and sudden change had taken place in the climate of that region." *

Dr. Rae lived some years on the banks of two of the great rivers of America, near to where they enter Hudson's Bay, and also on the Mackenzie River, which flows into the Arctic Sea, and had good opportunities of observing what takes place on these streams, all of which have large alluvial deposits, forming flats and shallows at their mouths.

Arctic America during Interglacial times.—We have seen that the eastern continent in Pleistocene times enjoyed in the Arctic regions interglacial conditions of climate. It is true that on the western continent we have not in Arctic regions such clear and satisfactory evidence of an interglacial period. But it would be rash to infer from this that the western continent was, in this respect, less favoured than the eastern. That we should find less evidence at the present day of former interglacial periods in Arctic America than in Arctic Asia, is what is to be expected, for the glaciation which succeeded interglacial periods has been far more severe in the former region than in the latter. The remains of the mammoth have, however, been found in Arctic America, in ice-cliffs at Kotzebue Sound, under conditions exactly similar to those of Siberia.

* "Phil. Mag." July, 1874, p. 60.

In Bank's Land, Prince Patrick's Island, and Melville Island, as in Northern Siberia, full-grown trees have been found in abundance at considerable distances in the interior, and at elevations of two or three hundred feet above sea-level. The bark on many of them was in a perfect state. Capt. McClure, Capt. Osborn, and Lieut. Mecham, by whom they were found, all agreed in thinking that they grew in the place where they were found.

It is true that more recent Arctic voyagers have come to the conclusion that these trees must have been drifted down the rivers from the south. There can be little doubt that the greater part of the wood found there, as in Siberia, is drift-wood. But may there not be also, as in Siberia, two kinds of wood?—a "Noashina" and an "Adamshina,"—a kind which was drifted and another kind which grew on the spot. This is a point which will require to be determined.

That so little has as yet been done in the way of searching for such evidence of interglacial periods is, doubtless, in a great measure due to the fact that most of those, if not all, who have visited those regions entertained the belief that there is an *à priori* improbability that a condition of climate which would have allowed the growth of trees in such a place prevailed so recently as Post-tertiary times. Even supposing those Arctic voyagers had considered the finding of interglacial deposits a likely thing, and had in addition made special search for them, the simple fact that they should have failed to find any trace of them could not, as we have already shown, be regarded as even presumptive evidence that none existed. Take Scotland as an example. Abundant relics of interglacial age have there been found from time to time; but amongst the many geologists who visit that country year by year, how few of them have the good fortune to discover a single relic. In fact a geologist might search for months, and yet fail to meet with an interglacial deposit. The reason is obvious. The last ice-sheet, under which Scotland was buried, was so enormous as to remove every remnant of the preceding interglacial land-surface, except here and there in deep and sheltered hollows, or in spots where it may happen to have been protected from the grinding power of the ice by projecting rocks. But all those places are now so completely covered with boulder-clay and other deposits that it is only in the sinking of pits, quarries, in railway-cuttings, and other deep excavations that traces of them accidentally turn up. Now if it is so difficult to find in temperate regions, in a place like Scotland, interglacial remains, how much more difficult must it be to meet with them in Arctic regions where the destructive power of the ice must have been so much greater.

Something like indications of an interglacial period appear to have been found by Professor Nordenskjöld in Spitzbergen. "In the interior of Ice-fjord," he says, "and at several other places on the coast of Spitzbergen, one meets with indications either that the polar tracts were less completely covered with ice during the glacial era than is usually supposed, or that, in conformity with what has been observed in Switzerland, interglacial periods have also occurred in the polar regions. In some sandbeds not very much raised above the level of the sea, one may, in fact, find the large shells of a mussel (*Mytilus edulis*) still living in the waters encircling the Scandinavian coast. It is now no longer found in the sea around Spitzbergen, having been probably routed out by the ice-masses constantly driven by the ocean-currents along the coasts."*

This testimony is the more valuable as it is given by an experienced geologist so much opposed to the theory of interglacial periods. A more special and thorough search of those beds might probably reveal further indications of interglacial age.

Was Greenland free from Ice during any of the Interglacial Periods?—There is nothing whatever improbable in the supposition that, during some of the earlier interglacial periods, when the eccentricity was about a maximum, the ice might have completely disappeared from Greenland, and the country become covered with vegetation.

Mr. Wallace thinks that the existence at present of an ice-sheet on Greenland is to be explained only by the fact that cold currents from the polar area flow down both sides of that continent. He further thinks that could these two Arctic currents be diverted from Greenland, "that country would become free from ice, and might even be completely forest-clad and habitable." †

I am inclined to agree with Mr. Wallace in thinking that the withdrawal of the two cold currents in question would effectually remove the ice. We know that Greenland is at present buried under ice, as has been shown on former occasions, simply because there happens to be about two inches more of ice annually formed than is actually melted. It certainly would not require any very great change in the present physical and climatic conditions of things to melt two

* "On Former Climate of Polar Regions," "Geol. Mag.," Nov., 1875, p. 531. See also "Geology of Spitzbergen," "Geol. Mag.," 1876, p. 267.

† "Island Life," p. 149.

additional inches per annum. If this were done, the ice would ultimately disappear. A simple decrease in the volume of the two currents might possibly bring about such a result. A cause more effectual would, however, be an increase in the temperature and volume of the Arctic branch of the Gulf Stream.

CHAPTER XII.

THE DISTRIBUTION OF FLORA AND FAUNA IN ARCTIC REGIONS.

Flora and Fauna of Iceland and the Faröe Islands destroyed by last Ice-sheet.—How was the present Flora and Fauna of these Islands introduced?—Professor J. Geikie's Explanation.—Ice and Ocean Currents as Transporting Agents.

WE have already seen (Chap. VIII.) that the ice-sheet which covered Scotland, Scandinavia, the Orkney and Shetland Islands, and the Outer Hebrides, towards the close of the Glacial period, was hardly less thick than that which mantled them at the climax of glacial cold. It is therefore evident that the flora and fauna of Greenland, Iceland, and the Faröe Islands could not possibly have survived in such highly glaciated conditions. The conclusion is thus forced upon us that the present flora and fauna of these places must have been introduced during Postglacial times. The question then arises, How are we to account for the introduction of the present flora and fauna? Professsor J. Geikie thinks* that, in order to account for the present life forms of these places, we must necessarily assume that, at some period during early postglacial times, a land connection must have existed between Greenland, Iceland, and the Faröe Islands and North-west Europe. But are we really obliged to assume a land connection? I am strongly impressed with the conviction that, in

* "Prehistoric Europe," p. 568.

respect to the Arctic regions, due attention has not been bestowed on floating ice and ocean currents, particularly the Gulf Stream, as transporting agents. Of the multitude of Scandinavian plants carried down by streams, land-slips, and other causes into the sea, some could not fail to find their way to Greenland, Iceland, and the Faröes. The Gulf Stream, of course, would not directly convey the plants from Scandinavia to Iceland and the Faröe Islands, but the return currents might. The water flowing out of Arctic seas must always equal in amount that flowing into them. The wide-extending Gulf Stream, to the north-west of Scandinavia, is met by an immense flow of polar water from the north, which polar current, on meeting the warm stream, passes underneath it, and continues its course southwards as an undercurrent.* Were the volume of the Gulf Stream considerably reduced, a thing which it evidently was during at least a part of the postglacial period, the polar current would not, as now, pass under the stream, but would pursue its course as a surface current outside of it in the direction of Iceland and the Faröes. This current would doubtless, now and again, carry to these places materials derived from the Gulf Stream. There can be no absolute separation between the Gulf Stream and the return currents. The water flowing northward warm ultimately returns cold. The two sets of currents are but parts of one general system of circulation.

As for Greenland, we have no grounds for concluding that the waters of the Gulf Stream do not actually reach its eastern shores, although by that time they may have become part of the cold return current. It is certainly not unlikely that, at a period when the Arctic seas were less encumbered by ice than at

* 'Climate and Time,' p. 219.

present, Scandinavian flora might be carried by the Gulf Stream to Greenland.

Ocean currents will not, of course, explain the migration of land animals, but floating ice may. During part of the postglacial period, when the volume of warm water passing along the Scandinavian shores was much less than at present, the sea would doubtless be frozen during winter. The result would likely be that some of the animals which might happen to get on the ice in spring, or in the early summer, when it broke up, would become entrapped and carried away on the floating rafts. The same system of currents which carried the flora to the shores of Iceland and the Faröe Islands would also carry to the same place many of the floating rafts. Most animals would survive for a week or two without food, and certainly none would perish on an ice-raft for want of fresh water. In addition, some of the animals might cross over to the Faröes during winter before the ice broke up. To some animals, an ice-bridge would be nearly as suitable as a land connection.

In order that those places should have become possessed of a flora and fauna, it was by no means necessary that there should have been a continuous transference of plants and animals from Scandinavia. All that was necessary was simply the introduction of a few members of each species; and this could hardly fail to have incidentally taken place during the course of a few centuries by the agencies which I have been detailing.

If a land connection, demanding an elevation of some 2000 or 3000 feet, be necessary in order to refurnish those places with a flora and fauna after a period of glaciation, then we must assume that there were prior elevations to that extent, or else assume that during interglacial periods, when those places would be in a condition even more favourable for life than they are at present, they must have been utterly devoid of life, without either plant or animal. This is evident; for the first period of extreme glaciation would as effectually destroy all life in those regions as did the last, and an elevation sufficient to produce a land connection would be as necessary after the first glaciation as it was after the last. The observations of Professor J. Geikie and M. Helland show that the Faröe Islands must have been separated from Scandinavia during the time of the last great extension of the ice by the same deep trough as now exists: so that, if the Faröe Islands were not totally devoid of all life during interglacial periods, there must, according to the land-connection theory, have been one or more elevations prior to postglacial times sufficient to unite those islands with the mainland.

Should it yet be found that Iceland and the Faröes contain interglacial beds with organic remains other than marine, this will so far militate against the theory of a land-connection; for it would prove that there must have been, at least, two immense elevations and subsidences of those regions since the beginning of the Glacial Epoch. Had these consisted simply of oscillations of sea level, depending in some way on the appearance and disappearance of the ice during the glacial and interglacial periods, a repetition of these oscillations is what might have been expected; but it is different if we suppose that they were simply incidental upheavals and subsidences of the land wholly unconnected with glacial phenomena. There is, in this case, an *à priori* improbability of even two such changes occurring in the same place since the beginning of the glacial epoch.

Some may be disposed to ask, Why did not ocean currents carry the flora and fauna of North America into Greenland, Iceland, and the Faröe Islands as readily as the plants and animals of Scandinavia? A very cursory glance at the path of ocean currents will show the improbability of landing on these coasts any forms of life floated from the North American continent.

CHAPTER XIII.

PHYSICAL CONDITIONS OF THE ANTARCTIC ICE-SHEET.

Sir Wyville Thomson on the Antarctic ice.—Testimony of iceberg.—Temperature of the Antarctic ice.—Heat derived from beneath.—Heat derived from the upper surface.—Heat derived from work by compression and friction.—Temperature of the ice determined by the temperature of the surface.—Temperature of the ice in some regions determined by pressure.

THERE are few subjects on which a greater amount of misconception as well as diversity of opinion, prevails, than in regard to the physical conditions of continental ice. This is more particularly true in reference to those physical and mechanical principles which limit the thickness of the ice-sheet and determine its form and mode of motion. These misapprehensions arise, in part at least, from an attempt to explain the conditions of continental ice on the principles of ordinary glaciers by geologists, who forget that between a continental ice-sheet and an ordinary glacier there is but little analogy.

At the present day, the only continental ice on the globe, with the exception of that of Greenland, is of course in the Antarctic regions. Here we have an ice-sheet rivalling in magnitude those on the northern hemisphere, even during the height of the Glacial Epoch. I know of nothing which will better illustrate the physical principles of continental ice than an attempt to answer the question, What is the probable thickness of the Antarctic ice-sheet? I shall therefore proceed to the consideration of this question.

In 'Climate and Time,' Chapter XXIII., I have endeavoured to show that the thickness of the ice on the Antarctic continent must be far greater than is generally supposed; that whatever be its thickness at the edge of the continent where it breaks up into bergs, that at the Pole or centre of dispersion it must be of enormous thickness.

Since the publication of my work, however, Sir Wyville Thomson, Director of the Scientific Staff of the *Challenger* Expedition, in a lecture on the condition of the Antarctic regions delivered at Glasgow some years ago, has come to a totally different conclusion. His conclusion is based chiefly on considerations relating to the principle of regelation and the physical nature of ice; and, as the same opinion is held by many, I shall examine it at some length. The following quotation from Sir Wyville's lecture will show his views on the subject:—

"There is one point in connection with the structure of icebergs which is of great interest, but with regard to which I do not feel in a position to form a definite judgment. It lies, however, especially within the province of a distinguished professor in the University of Glasgow, Dr. James Thomson, and I hope he will find leisure to bring that knowledge to bear upon it which has already thrown so much light upon some of the more obscure phenomena of ice. I have mentioned the gradual diminution in thickness of the strata of ice in a berg from the top of the berg downwards. The regularity of this diminution leaves it almost without a doubt that the layers observed are in the same category, and that therefore the diminution is due to subsequent pressure or other action upon a series of beds which were at the time of their deposition pretty nearly equally thick.

About 60 or 80 feet from the top of an iceberg the strata of ice, a foot or so in thickness, although of a white colour, and thus indicating that they contain a quantity of air and that the particles of ice are not in close apposition, are still very hard, and the specific gravity of the ice is not very much lower than that of layers not more than 3 inches thick nearer the water-line of the berg. Now, it seems to me that this reduction cannot be due to compression alone, and that a portion of the substance of these lower layers must have been removed.

"It is not easy to see why the temperature of the earth's crust, under a widely extended and practically permanent ice-sheet of great thickness, should ever fall below the freezing-point, and it is a matter of observation that at all seasons of the year vast rivers of muddy water flow into the frozen sea, from beneath the great glaciers which are the issues of the ice-sheet of Greenland. Ice is a very bad conductor, so that the cold of winter cannot penetrate to any great depth into the mass. The normal temperature of the crust of the earth at any point where it is uninfluenced by cyclical changes is, at all events, above the freezing-point, so that the temperature of the floor of the ice-sheet would certainly have no tendency to fall below that of the stream which was passing over it. The pressure upon the deeper beds of the ice must be enormous; at the bottom of an ice-sheet 1400 feet in thickness it cannot be much less than a quarter of a ton on the square inch. It seems therefore probable that, under the pressure to which the body of ice is subjected, a constant system of melting and regelation may be taking place, the water passing down by gravitation from layer to layer until it reaches the floor of the ice-sheet, and finally working out channels for itself between the ice and the land, whether the latter be sub-aërial or submerged.

"I should think it probable that this process, or some modification of it, may be the provision by which the indefinite accumulation of ice over the vast nearly level regions of the 'Antarctic Continent' is prevented, and the uniformity in the thickness of the ice-sheet is maintained; that, in fact, ice at the temperature at which it is in contact with the surface of the earth's crust within the Antarctic regions, cannot support a column of itself more than 1400 feet high without melting."*

The subject is one of very considerable importance, not merely in relation to the Antarctic regions at the present day, but also in its bearings on the condition of things generally during the Glacial Epoch. For if Sir Wyville Thomson's conclusions in reference to the thickness of the Antarctic ice be true, they must hold equally true for the ice of the Glacial Epoch, and consequently would modify to a large extent prevailing conceptions regarding the physical condition of our country during that epoch.

They are therefore conclusions worthy of discussion, and, as they are diametrically opposed to those arrived at by myself, I have thought of considering the subject in somewhat fuller detail, the more so as new elements in the question have since been introduced.

At the very outset of the inquiry it must be observed that the question of the thickness of the ice covering the Antarctic continent is one which cannot be determined by direct observation. No one, as yet, has ever been able to set his foot on that continent; and the perpendicular wall forming the outer edge of its icy mantle is nearly all that has been seen of it. Direct measurements, and some other facts to which we shall shortly refer, show with tolerable certainty what is the probable average thickness of the ice-sheet at its outer circumference; but observation can tell us nothing whatever about the thickness of the ice in the interior, which is the question at issue. This has to be

* "Condition of the Antarctic Region," p. 23. "Nature," vol. xv., p. 122.

determined by purely physical and mechanical considerations, based, it is true, on data derived from observation.

It fortunately happens, however, that the very circumstances that render the region so difficult to get at are those which at the same time tend to simplify the problem. The Antarctic region is the most inaccessible on the globe, but of all regions it is the one where the physical conditions are most uniform and least under the influence of contingent circumstances, such as those resulting from the presence of warm ocean currents in one place and cold currents in another, or of great masses of land in one part and an open sea in another. We have not in the Antarctic, as in the Arctic region, well-marked warm and moist aërial currents and cold and dry winds blowing athwart different areas. Surrounding the South Polar continent lies an unbroken ocean, in an almost uniform climatic condition. This region also, as Sir Wyville Thomson remarks, is "continuously solid,—that is to say, it is either continuous land or dismembered land fused into the continental form by a continuous ice-sheet." In this case we can treat it as one continuous continent. The South Pole being safely assumed to be in the centre of the sheet, we have here what we perhaps never had on the northern hemisphere, even during the Glacial Epoch—a polar *ice-cap.* We have the pole in the centre of the cap; therefore, at equal distances from the pole or centre, the conditions in every respect, both as to climate and the thickness of the ice, may be assumed to be the same, for no reason can be assigned for supposing the conditions in separate areas upon the same parallel of latitude to differ. Thus, as a purely physical and mechanical problem, the conditions could hardly be more simplified.

Testimony of Icebergs.—We shall now enter into the consideration of the question. In the first place, the conclusion that "ice at the temperature at which it is in contact with the surface of the earth's crust within the Antarctic regions cannot support a column of itself more than 1400 feet high without melting" is in direct opposition to known facts.

The immense tabular icebergs found in the Southern Ocean, which have been so well described by Sir Wyville Thomson, are, of course, portions broken off the edge of the ice-sheet, and the thickness of the bergs represents the thickness of the ice-sheet at the place where they broke off. Now, some of these icebergs have been found to be more than three times thicker than the limit assigned by Sir Wyville. The following are a few out of the many examples which might be adduced of enormous icebergs, taken from the Twelfth Number of the "Meteorological Papers," and from the excellent paper of Mr. Towson on the "Icebergs of the Southern Ocean," both published by the Board of Trade:—

Sept. 10th, 1856.—The "Lightning," when in lat. 55° 33′ S., long. 140° W., met with an iceberg 420 feet high.

Nov., 1839.—In lat. 41° S., long. 87° 30′ E., numerous icebergs 400 feet high were met with.

Sept., 1840.—In lat. 37° S., long. 15° E., an iceberg 1000 feet long and 400 feet high was met with.

Feb., 1860.—Captain Clark, of the "Lightning," when in lat. 55° 20′ S., long. 122° 45′ W., found an iceberg 500 feet high and 3 miles long.

Dec. 1st, 1859.—An iceberg, 580 feet high and from 2½ to 3 miles long, was seen by Captain Smithers, of the "Edmond," in lat. 50° 52′ S., long. 43° 58′ W. So strongly did this iceberg resemble land that Captain Smithers believed it to be an island, and reported it as such, but there is little or no doubt that it was in reality an iceberg. There were pieces of drift-ice under its lee.

Nov., 1856.—Three large icebergs, 500 feet high, were found in lat. 41° S., long. 42° E.

Jan., 1861.—Five icebergs, one 500 feet high, were met with in lat. 55° 46′ S., long. 155° 56′ W.

Jan., 1861.—In lat. 56° 10′ S., long. 160° W., an iceberg 500 feet high and half-a-mile long was found.

Jan., 1867.—The barque "Scout," from the West Coast of America, on her way to Liverpool, passed some icebergs 600 feet in height and of great length.

April, 1864.—The "Royal Standard" came in collision with an iceberg 600 feet in height.

Dec., 1856.—Four large icebergs, one of them 700 feet high and another 500 feet, were met with in lat. 50° 14 S., long. 42° 54′ E.

Dec. 25th., 1861.—The "Queen of Nations" fell in with an iceberg in lat. 53° 45′ S., long. 170° W., 720 feet high.

Dec., 1856.—Captain P. Wakem, ship "Ellen Radford," found, in lat. 52° 31′ S., long. 43° 43′ W., two large icebergs, one at least 800 feet high.

Mr. Towson states that one of our most celebrated and talented naval surveyors informed him that he had seen icebergs in the southern regions 800 feet high.

March 23rd, 1855.—The "Agneta" passed an iceberg in lat. 53° 14′ S., long. 14° 41′ E., 960 feet in height.

Aug. 16th, 1840.—The Dutch ship "General Baron Von Geen" passed an iceberg 1000 feet high in lat. 37° 32′ S., long. 14° 10′ E.

From the fact that the ice forming the upper layers of the icebergs is less dense than that of ordinary ice, Sir Wyville Thomson estimates that as much as one-seventh part of the berg may be above water-line. But, for the following reasons, I am unable to agree with this estimate. It is true, as he remarks, that the white ice which forms the upper portion of the berg is less dense than ordinary ice, being composed of recent snows; but, on the other hand, this will be counterbalanced by the greater density of the lower portions of the berg, which have been subjected for ages to enormous pressure. I hardly think that there is any good reason to conclude that the *mean* density of the bergs is much under that of ordinary ice, namely, 0·92.*

But even if we admit that as much as one-seventh of the berg is above water, still a berg 500 feet in height would be 3500 feet in thickness, and one 600 feet would be 4200 feet thick, while one 720 feet high, of the tabular form, would be 5040 feet, or nearly a mile in thickness.

It would not, of course, be safe to conclude that the thickness of the ice below water bears always the same proportion to the height above. If the berg, for example, be much broader at its base than at its top, the thickness of the ice below water would bear a less proportion than that indicated by the difference of

* It is true that, from observations made ("Quart. Journ. Geol. Soc., Feb., 1877) on the density of ice in Disco Bay, Mr. Amund Helland found that, in consequence of the amount of air-bubbles contained in the ice, its density was only 0·886, and from this he concluded that one-seventh of the bergs was above water. But he does not state at what part of the berg his specimens were taken. If they were taken from near the top, or even at the water-line, it might have been expected that the density would be very considerably under that of ordinary ice.

specific gravity of ice and water. But a berg such as that recorded by Captain Clark, 500 feet high and 3 miles long, may be relied upon as having the proportionate thickness under water. The same may be said of the one seen by Captain Smithers, which was 580 feet high, and so large that it was taken for an island.

It may be here remarked that a berg does not stand higher out of the water because the lightest side happens to be uppermost. The height above water is determined by the *mean* density of the berg, and is the same no matter how the various densities may be distributed through the mass. It would be the same though the berg were turned upside down. This follows as a necessary consequence from the fact that the amount of water displaced by the berg is equal to its weight, and of course it is the same whatever side be uppermost.

To evade the force of the evidence derived from the testimony of the icebergs, it is asserted by some that the heights thus recorded are mere guesses, and not the result of actual measurement. But such an opinion is in direct contradiction to the express declaration of Admiral Fitzroy, who collected the evidence on the subject. He states that "by angular and reliable measurements some of them have been found to be six or eight hundred feet high and several miles in circumference."

But more than this, if Captain Smithers, for example, did not actually measure the iceberg to which we have referred, he could not have known that its height was 580 rather than 600 feet. The very fact that he stated it to be 580, and not 500 or 600, or even 550 feet, surely implies that he really measured it. The assertion that a person is 5 feet 6 inches or 5 feet 8 inches in height would not imply that this was his measured height; but if we asserted that he was 5 feet 6½ inches or 5 feet 8¾ inches high, we should necessarily convey this impression. In like manner, Captain Smithers, by assigning to the iceberg an altitude so particular as that of 580 feet, distinctly conveys the impression that such was the height obtained by actual measurement. Similarly we conclude that the captains of the "Queen of Nations" and the "Agneta" actually measured the icebergs which they respectively declared to be 720 and 960 feet high.

In reply, it may perhaps be asserted that no record of an iceberg 500 or 600 feet in height is to be found in the log-books of the Navy, and that all those instances of enormous icebergs have been given by masters of merchant vessels, who, as a rule, are not so competent to make accurate measurements. It is doubtless true that the latter generally are not so well qualified for such work as naval officers; but it is hardly credible that they should all have gone so far astray in their measurements as to estimate heights at 500 and 600 feet which in reality were only 200 feet. Now, if but one berg 500 feet high has ever been seen in the Southern Ocean, it is proved that even twice 1400 feet is not the limit of the thickness of the Antarctic ice.

But it is not the case that no naval officer has met an iceberg of those enormous dimensions; for, as we have already seen, Mr. Towson states that one of our most celebrated and talented naval surveyors informed him that he had met icebergs in the southern regions 800 feet high. It is, however, not to be wondered at that so few naval officers have seen such bergs, for they are of very rare occurrence. They have been met with chiefly in latitudes that are traversed by thousands of merchant ships for one vessel belonging to the Navy. And perhaps not one out of every ten thousand merchantmen has ever fallen in with one of the great ice islands we now speak of.

The testimony from icebergs may therefore be regarded as decisive against the opinion that the Antarctic ice cannot be more than 1400 feet thick.

That 1000 or 2000 feet cannot be the limits of thickness attained by continental ice is amply proved by the geological evidence, which goes to show that during the Glacial Epoch the ice in some places much exceeded 1400 feet. Professor Dana has proved that during the period in question the thickness of the ice on the American continent must in many places have been considerably above a mile. He has shown, that over the northern border of New England, the ice had a mean thickness of 6500 feet, while its mean thickness over the Canada water-shed, between St. Lawrence and Hudson's Bay, was not less than 12,000 feet, or upwards of 2¼ miles.* Professor Erdmann and Mr. Amund Helland have shown that the ice in some parts of Scandinavia was at least 6000 feet thick. It has been proved by M. Guyot and others, that the great valley of Switzerland was formerly filled with a mass of ice between 2000 and 3000 feet in thickness. Mr. Jamieson found that the isolated mountain of Schiehallion, in Perthshire, 3500 feet high, is marked near its top as well as on its flanks, and this not by ice flowing down the side of the hill itself, but by ice passing over it from the north. Dr. James Geikie has shown † that the ice between the mainland and the Outer Hebrides was as much as 3700 feet in thickness. The great mass of ice from Scandinavia,

* See "American Journal of Science and Arts," March, 1873.

† "Quart. Journ. Geol. Soc.," vol. xxxiv., p. 861.

filling the Baltic and the North Sea, during the Glacial Epoch must have been over 3000 feet thick at least.*

The Temperature of the Antarctic Ice.

In examining the physical reasons which have been advanced for the limit assigned to the thickness of the Antarctic ice-cap, we must first consider the probable temperature of the ice; for not only does the thickness of the sheet depend, as we shall see, to a considerable extent on the temperature of the ice, but misapprehensions on this point will tend to vitiate all our reasoning on the subject.

There are but three directions from which the ice-cap can receive an appreciable amount of heat, viz., (1) the air above; (2) the earth beneath; and (3) the work of compression. Other sources can yield little, if any at all. For instance, the amount carried inward horizontally from the outer edge of the cap by conduction must be infinitesimal, and indeed can never affect the interior, as the ice moves outward more rapidly than the heat can possibly travel inward.

Heat derived from Beneath.—We shall begin with the consideration of the heat derived by the bottom of the sheet from the earth's crust. The researches of Sir William Thomson enable us to determine with a tolerable degree of certainty the amount received from this source. He tells us that through every square metre of the earth's surface 220 metre-tons, or 1,613,700 foot-pounds, of underground heat pass upwards annually. Through every square foot, therefore, there must come 149,600 foot-pounds. This amount is sufficient to melt a layer of ice, already at the melting-point, one-fifth of an inch in thickness. But under-

* 'Climate and Time,' Chap. xxvii.

ground heat would probably be insufficient to melt even so small a layer, since a portion of the heat must, doubtless, be expended in passing through and maintaining at the melting-point a few inches of the ice at the bottom of the sheet.

At first we might be apt to suppose that underground heat ought to travel up through the ice in the same way as through the strata of the earth below, and to make its presence sensibly felt at no great distance from the surface of the sheet. This, however, is impossible; for (1) the greater part of the heat is spent not in raising the temperature, but in melting the ice; and (2) the ice when melted immediately runs off, carrying the heat along with it. But it will be replied, that, notwithstanding this, if the temperature of the ice be much below the freezing-point, the heat constantly passing into the solid layers at the bottom of the sheet, though trifling, ought in course of ages to pass up through the ice, affecting its temperature not for a few inches, as I have supposed, but for a thickness of a great number of feet. Were the ice, like the ground underneath on which it rests, to remain immovable, this would no doubt be the case; but the sheet is in a state of constant motion outwards from the centre of dispersion, and no sooner is a particle of the ice heated than it moves away, and its place is supplied by another particle from behind, which in turn requires to be heated. Besides, the ice has always a downward as well as a horizontal motion; for all the ice found at the bottom comes primarily from the top, and that removed from below is replaced from above. Hence not only is internal heat from below carried away by the horizontal flow of the ice, but the upward motion of the heat is checked by a downward flow of the ice from above; and the ice is, in all probability, moving downwards more rapidly than the heat is travelling upwards. We must therefore conclude that underground heat is confined to a very thin layer of the cap at the bottom, and that its effects, either in melting the ice or in raising its temperature, are so trifling that they may be practically disregarded in our present inquiry.

It must further be observed that when it is stated that underground heat will maintain at the melting-point the ice in contact with the ground, it is not meant that it will maintain it at the temperature of 32°, for, as Prof. James Thomson discovered, the temperature at which the ice melts is lowered by pressure at the rate of about 0·0137° F. for every atmosphere of pressure. In the present case the pressure depends upon the thickness of the ice; so that, if the sheet be 1400 feet deep, the melting-point will be 31°·5; if half-a-mile deep, it will be 31°; if 1 mile deep, 30°; and so on.*

Heat derived through the Upper Surface.—It follows, from what has already been shown, that the greater

* The melting-point does not, however, vary uniformly with the pressure; for Mousson (Ann Chim. et Phys., 3rd series, lvi., p. 257, 1859) found that it required a pressure of 13,000 atmospheres to lower the melting point to zero F., whereas, if the melting-point had decreased in proportion to the increase of pressure, a pressure of 2337 atmospheres would have been sufficient. Boussingault succeeded in lowering the melting-point 11° below zero F., but the amount of pressure employed was not determined (Ann. Chim. et Phys., xxvi., p. 544, 1872).

The fact that the melting-point of ice would be lowered by pressure, or rather that pressure would prevent freezing, was suggested nearly a century ago by Dr. Charles Hutton, Professor of Mathematics in the Military Academy of Woolwich. From certain experiments on the expansive force of ice, made in Canada by Major Williams, in the year 1784-85, Dr. Hutton makes the following remarks:—

"From these ingenious experiments we may draw several conclusions:—First. We hence observe the amazing force of the part, if not nearly all, the heat possessed by the ice must have been received through the upper, and not the under, surface of the sheet. But what we are at present concerned with is not so much the amount of heat received by the ice as the temperature at which the heat can maintain the ice. In short, the question to be determined is—What is the temperature of the Antarctic ice? Now, if nearly all the heat possessed by the ice has been received from the upper surface of the sheet, the temperature of the mass must be mainly determined by that of the surface, and cannot be far above the mean temperature of the surface. If so, the temperature of the ice must evidently be very considerably below the freezing-point.

(1.) If we suppose the heat to be transmitted from the surface downwards by *Conduction*, we must necessarily conclude that the surface is at a higher temperature than the ice below, for conduction can only take place from a hot to a colder body; and this process could not possibly maintain the mass of the ice below at a temperature equal to the mean temperature of the surface. The general tendency of conduction would, therefore, be to keep the ice beneath at a lower temperature than that at the surface.

(2.) The work of *Radiation*, however, would probably have the opposite tendency. The heat received by direct radiation from the sun could not possibly raise

expansion of the ice, or the water in the act of freezing; which is sufficient to overcome perhaps any resistance whatever; and the consequence seems to be, either that the water will freeze, and, by expanding, burst the containing body, be it ever so thick and strong; or else, if the resistance of the body exceeds the expansive force of the ice, or of water in the act of freezing, then, by preventing the expansion, *it will prevent the freezing, and the water will remain fluid, whatever the degree of cold may be.*" (Trans. Roy. Soc. Edin., vol. ii., p. 27).

the temperature of the ice above 32°, but the heat lost by radiation might lower the temperature to far more than 32° below zero. If the heat received from the sun's rays should keep the surface of the ice at, say, the melting-point during the summer, and the heat lost by radiation should keep the surface at, say, 50° below the melting-point during winter, which is not an extravagant supposition, the mean temperature of the surface would then be 25° below the melting-point, or 7° F. But the mean temperature of the underlying ice would not be so low; for the low mean temperature of the surface is almost wholly due to loss by radiation into stellar space during winter, and this loss would be chiefly confined to the surface. Had the surface been rock instead of ice, the rise of temperature during summer would have been about as great as the decrease during winter, and consequently the mean temperature would have been much higher than in the case of ice. Hence, the difference between the mean temperature of a rock surface and that of the rock below would not be so great as in the case of ice. The tendency of direct radiation, therefore, is to maintain the surface of the ice-sheet at a lower temperature than that of the underlying mass.

(3.) This tendency is strengthened by another circumstance which comes into operation. During summer, a large portion of the direct heat from the sun is spent in melting the surface ice. The melted ice passes down through crevasses and openings in the sheet, thus carrying the temperature along with it. The heat of summer is by this means carried down below the surface, but not so the cold of winter.

The melting of the ice on the Antarctic continent will be greatly retarded, however, by the coldness of the air, the temperature of which, even during summer, is considerably below the freezing-point. A wind, a few degrees below the freezing-point, blowing on the icy surface would probably re-freeze the ice as rapidly as the sun's rays could melt it. These conditions differ entirely from those that obtain in the Arctic regions. In the latter the air in summer is above the freezing-point, and consequently assists the sun in melting the ice, whereas in the Antarctic regions it is below the freezing-point, and tends to prevent the sun from melting the ice. This circumstance explains the fact, which so much surprised Sir James Ross, that no streams of water flow off the Antarctic ice, similar to those that escape from the great ice-fields of Greenland.

Such water as formed on the surface could not penetrate to any considerable depth, for the ice, as we shall presently see, being much below the freezing-point, the water would be re-frozen before it could descend to any great depth.

It therefore follows that the great mass of ice, up to within a short distance of the surface, can be very little affected by heat transmitted either by conduction, or by radiation, or by water from ice melted at the surface.

Heat derived from Work of Compression and Friction.—I shall now consider the third and last source from which the ice can obtain heat, viz., *Work of Compression and Friction.* We are fortunately able to come to a pretty definite conclusion in regard to the total amount of heat derivable from this source. The force employed is Gravity, and we can thus determine with certainty the greatest amount of work which that can possibly perform on the ice. Mere pressure, however great, cannot of course generate heat unless it perform work, and the heat thus generated is not proportionate to the pressure, but to the work performed, and the amount of work done by pressure is proportionate to the space through which the pressure continues to act. When the pressure is gravity, the work is measured by the distance that the body is allowed to descend. A pound weight descending 1 foot performs 1 foot-pound of work; descending 2 feet it performs 2 foot-pounds; and so on in proportion to the number of feet of descent. In estimating the total amount of work which gravity can perform in the descent of a glacier down the side of a mountain, we measure the work by the vertical distance the glacier descends; but in the case of the Antarctic ice-cap, the slope of the ground does not enter as an element into our calculations, for the ground is assumed to have no slope, the continent being regarded as flat. The surface no doubt may have great irregularities, such as hills and mountain-ridges; those irregularities, however, do not assist gravity, but rather act as obstructions to the general flow of the ice.

Nevertheless, just as in the case of a glacier, the amount of work that gravity can perform is determined by the distance the ice can descend; and this distance is determined not by the slope of the ground, but by the thickness of the sheet. If the Antarctic ice-sheet be 1400 feet in thickness, the greatest distance to which a pound of ice can descend is of course 1400 feet. Gravity acting on this pound of ice can, therefore, perform only 1400 foot-pounds of work. But, in order that gravity may do so, the pound must descend the whole distance from the surface to the bottom of the sheet. In estimating the total amount of heat which could possibly have been conferred on the ice by gravity, we must find the mean vertical distance to which the ice has descended. This, of course, in the present case, is equal to half the thickness of the sheet, viz., 700 feet. If 1400 feet, as Sir Wyville Thomson supposes, be the thickness of the Antarctic ice-cap, 700 foot-pounds per pound is the utmost quantity of work that gravity can have performed on the ice. Supposing the whole of this work had been employed in heating the ice by compression, or by the friction of the particles of the ice on one another, or on the rocky floor of the sheet, the heat generated would not have amounted to one thermal unit per pound of ice. The specific heat of ice being about one-half that of water, the total work of compression—assuming that it had all been converted into heat, and the heat equally distributed through the entire mass of the cap—would not have raised the temperature of the ice by 2°.

The foregoing considerations do not afford a means of determining what the actual temperature of the great mass of the ice below the surface is. They show, however, that whatever that temperature may be, it is not very materially affected either by the heat of compression, or by undergound heat, or by that transmitted from the surface either by conduction or by melted ice.

On what, then, does the temperature of the ice mainly depend?

Temperature of the Ice Determined by the Temperature of the Surface.—The temperature of the great mass of the ice is mainly determined by the mean temperature of the upper surface of the sheet. All the ice down to the bottom of the sheet originally came from the surface. It once existed at the surface in the form of a coating of snow, which, becoming consolidated into ice, was afterwards covered over with fresh layers of snow, while these in turn, passing into ice, were buried under succeeding snows, and so on. The ice that formed the surface a century ago now lies buried below the ice of a hundred years, and a hundred years hence its present position will be occupied by the surface ice of to-day. There is not only a constant motion of the ice from the pole outwards, but a constant downward motion as layer by layer is successively formed on the surface.

From what has been proved regarding the small quantity of heat which can be directly transmitted through the ice, it follows that the superficial layers will carry down with them pretty much the same temperature which they possessed at the surface at the time when they were covered up by succeeding snows. Any heat which they can derive from the work of compression, as has been shown, is but trifling. Heat transmitted by conduction could not possibly raise the temperature of the underlying ice above that of the surface; neither could the heat from direct radiation, nor that derived from melted ice.

As the temperature of the ice, then, cannot be much above the mean temperature of the surface, which is far below the freezing-point, it follows that the underlying mass must also be below the freezing-point. The very low temperature of the superficial layers is due to the fact that the mean temperature of the air above the surface is far below the freezing-point—a temperature which the icy surface cannot much exceed. The sun during summer may possibly heat the air sometimes above the freezing-point, but it cannot, of course, so raise the temperature of the ice without melting it.

Again, as solid ice is a better radiator than gaseous air, the surface of the sheet during winter would probably have its temperature lowered by radiation to a greater extent than the air. The probability is that the mean annual temperature of the surface is as low as, if not lower than, that of the air over it. And although the mean temperature of the regions around the South Pole has not been ascertained by direct observation, yet it certainly cannot be much higher than that of those around the North Pole, which we know is but a few degrees above zero F.

Now, if the mean annual temperature of the air over the Antarctic ice-sheet be not very much above zero F., then that of the surface of the sheet cannot be much higher; and if this be so, it follows, from what has been already advanced, that the temperature of the great mass of the ice down to near the bottom of the sheet must be considerably below the freezing-point.

Temperature of the Ice in some Regions determined by Pressure.—In regions such as Switzerland, where the mean temperature is above the freezing-point, the temperature of the ice in the interior of a glacier is not determined by the air above. The tendency of the air in this case is to keep the entire mass of the glacier at the melting-point. But as the temperature of the melting-point depends upon the pressure, the temperature of the glacier at any depth from the surface will depend upon the pressure to which the ice at that depth is subjected. At and near the surface, where the pressure is small, the temperature of the ice will be 32°. At the depth of a quarter of a mile, where the pressure would be equal to about 36 atmospheres, the temperature would be, not 32°, as at the surface, but 31°·5. And if the glacier were half a mile thick the temperature at the bottom would be 31°, and so on in proportion to the thickness of the glacier. This lowering of the melting-point could not, however, go on without limit, for in a country like Switzerland a point would soon be reached where the ice could no longer retain the solid form. If, for example, owing to the heat of the climate, we could not have ice at a lower temperature than say 30°, then a glacier over 1 mile in thickness would be an impossibility, for the bottom of a glacier of greater thickness would not remain solid at that temperature.

Having considered the various circumstances affecting the temperature of the Antarctic ice, and the sources from which it derives its heat, we have found that the temperature of the ice must be considerably under the freezing-point. We are now prepared to examine the reasons which have been adduced for concluding that about 1400 feet is the probable limit to the thickness of the Antarctic ice-sheet.

CHAPTER XIV.

PHYSICAL CONDITIONS OF THE ANTARCTIC ICE-SHEET.
—Continued.

Limit to Thickness of the Ice resulting from Melting produced by Pressure.—Supposed Diminution in Thickness of Ice-strata from Compression and Melting.—Centre of Dispersion.—Ice thickest at Centre of Dispersion.—Thickness at Pole independent of Amount of Snowfall.—Rate of Motion of the Antarctic Ice.—Probable Thickness at the Pole.—Ice of the Glacial Epoch.

Examination of the Reasons for Supposing that the Antarctic Ice-cap is not of Enormous Thickness.

1. *Limit to the Thickness of the Ice resulting from Melting produced by Pressure.*—Pressure will produce a melting of the ice in two totally different ways, viz., either by lowering the melting-point or by the work of compression. I shall consider both cases, and see if any ground is afforded by either in support of the conclusion that the Antarctic ice can be only one or two thousand feet in thickness.

(a.) Melting produced by the Lowering of the Melting-point.—The pressure exerted by a column of ice of half a mile in height would, as we have seen, lower the melting-point 1°; consequently, if the column were at the temperature of 32°, its base, being 1° above the melting-point, would not remain in the solid condition. To prevent the ice melting, the temperature of the base would require to be as low as 31°. Therefore, if 32° were the temperature of the Antarctic

ice, Sir Wyville Thomson's conclusion, that the sheet cannot be more than 1400 feet in thickness, would follow as a matter of course.

But his supposition that, owing to internal heat coming through the earth's crust, the bottom of the Antarctic ice-sheet is kept at the temperature of 32°, cannot be sustained. It is this fundamental error, as I conceive it to be, which has led Sir Wyville astray, and induced him to believe that the ice cannot be of excessive thickness.

"The normal temperature of the crust of the earth," says Sir Wyville, "at any point when it is uninfluenced by cyclical changes, is, at all events, above the freezing-point, so that the temperature of the floor of the ice-sheet would certainly have no tendency to fall below that of the stream which was passing over it. . . . In fact, ice at the temperature at which it is in contact with the surface of the earth's crust within the Antarctic regions, cannot support a column of itself more than 1400 feet without melting."

In the question under consideration we are directly concerned with the temperature of the *surface* of the earth's crust only,—the floor on which the ice-sheet rests,—and not with the temperature below the surface. It is perfectly true that at considerable depths below the surface internal heat maintains a temperature above the freezing-point; but it is not true that it determines the heat of the surface. Underground heat produces scarcely any sensible influence on the temperature of the surface, which is determined almost wholly by that of the air and other external agencies. The temperature, in short, is determined from *above*, and not from *beneath*. In warm countries, where the temperature of the air is high, that of the surface is high also. And so likewise in cold climates, the low temperature of the air gives a comparatively low temperature to the surface. Suppose our globe to be enveloped for some thousands of years with a covering at the uniform temperature of say 100°; and suppose, further, that 5000° should represent the temperature of the earth's mass; then, in such a case, there would be a gradual decrease of temperature from 5000° at the centre to 100° at the surface. Let us suppose now the warm covering is removed, and replaced by one at –100°. In the course of some thousands of years there will be a gradual decrease of temperature from 5000° at the centre, as before, to –100° at the surface. *Internal* heat limits the temperature at the centre, but *external* heat limits in every case the temperature at the surface.

To maintain, as Sir Wyville Thomson does, that 32° is the temperature of the floor on which the Antarctic ice-sheet rests, is virtually to beg the whole question at issue. It is the temperature of the ice that determines that of the floor on which it rests, and not the latter that determines the former.

What the temperature of the ground under the Antarctic ice-sheet may be is a question which at present we have no means of ascertaining with certainty; we know only that it must be far below the freezing-point, for the ice resting on it is considerably under that point.

Although the temperature of the ice must impose a limit to the thickness of the sheet, underground temperature cannot do so, for the temperature of the ice is not determined by underground heat.

But supposing we knew the temperature of the Antarctic ice, yet this knowledge would not enable us to determine with certainty the limit imposed by temperature on the thickness of the ice. For, excepting in cases where the temperature is but very little below 32°, we are at present, in the absence of experiments, unable to say what would be the amount of pressure necessary to lower the melting-point to any assigned temperature. The experiment of Mousson shows that Professor Thomson's formula, $t = 0{\cdot}0137^{\circ} n$, does not hold true when the pressure is excessively great. A pressure of 73 atmospheres will lower the melting-point from 32° to 31°; but if Mousson's experiment is to be depended upon, a pressure of 400 atmospheres would be necessary to lower the melting-point from 1° to 0°. That is to say, were sufficient pressure applied to lower the melting-point to 1°, it would require an additional 400 atmospheres to lower it to 0°. The *rate* at which the melting-point is lowered by pressure is evidently not uniform, but decreases with the increase of pressure. Were the temperature of the ice at the South Pole as low as 32° below the frezing-point, which doubtless it is not, it would, according to Mousson's experiment, support a thickness of not less than 90 miles. But if the rate did not diminish with the pressure, but remained uniform, a pressure of 16 miles would be the limit.

From what has already been proved, I think we may safely assume that the ice at the South Pole may be at least ten or twelve degrees below the freezing-point. We are unable to say what thickness of ice this temperature could support, but we know that it must be over 6 but under 30 miles.

But whatever the actual temperature of the Antarctic ice may be, if the sheet be as thick as the temperature will admit, then underground heat can never raise the temperature of the surface under the sheet sensibly above that of the ice. This is evident, because it cannot raise the temperature of the ice above the melting-point corresponding to the pressure, and the ice will always keep the floor at sensibly the same temperature as itself. In short, in determining the thickness of the Antarctic ice, underground heat does not enter as an element into our calculations, and, so far as the melting of the ice produced by the lowering of the melting-point is concerned, the Antarctic ice at the Pole may be a dozen of miles in thickness as readily as 1400 feet.

(b.) Melting produced by Work of Compression and Friction.—"The pressure upon the deeper beds of ice," says Sir Wyville Thomson, "must be enormous; at the bottom of an ice-sheet 1400 feet in thickness it cannot be much less than a quarter of a ton on the square inch. It seems, therefore, probable that, under the pressure to which the body of ice is subjected, a constant system of melting and regelation may be taking place, the water passing down by gravitation from layer to layer until it reaches the floor of the ice-sheet, and finally working out channels for itself between the ice and the land, whether the latter be subaërial or submerged."

As has already been stated, no amount of pressure, however great, has the least tendency whatever to produce a melting of the ice by heat unless this pressure performs work, and the quantity of ice melted will then be, not in proportion to the pressure, but to the work performed by the pressure. The pressure here referred to, which is supposed to produce the melting, is the *weight* of the ice, or, in other words, the force of gravity.

When considering the amount of heat derived from work of compression, it was proved that, in the case of the Antarctic ice-sheet, the total amount of work which can possibly be performed by gravity is determined by the thickness of the sheet. It was shown that, if 1400 feet be the thickness of the sheet, 700 foot-pounds per pound of the sheet is the greatest amount of work that gravity can perform. It follows therefore that, supposing the whole of the work is employed in heating the ice by compression and friction, the heat thus generated would amount to only 0·9 of a thermal unit per pound of ice. It must be obvious that, in the case of a flat and tolerably uniform sheet like the Antarctic, in which the pressure must of course be pretty evenly distributed, little or no melting can take place from this cause, as it requires not 0·9 of a thermal unit, but 142 thermal units, to melt a pound of ice already at the melting-point. The total work of 158 pounds of ice would need to be concentrated upon one pound in order to melt it. But such an unequal distribution of force in a sheet so uniform is at least extremely improbable. The tabular form of the southern icebergs, with their stratification parallel to their upper surface, shows the flat character of the ground on which they have been formed. This circumstance appears to have particularly struck Sir Wyville Thomson, as well as all who have visited the Antarctic regions. "The stratification," says Sir Wyville, "in all the icebergs which we saw, was, I believe, originally horizontal and conformable, or very nearly so. I never saw a single instance of deviation from the horizontal and symmetrical stratification which could in any way be referred to original structure. . . . . As I have already said," he continues, "there was not, so far as we could see, in any iceberg, the slightest trace of structure stamped upon the ice in passing down a valley, or during its progress over *roches montonnées,* or any other form of uneven land; the only structure, except the parallel stratification, which we ever observed which could be regarded as bearing upon the mode of original formation of the ice-mass, was an occasional local thinning out of some of the layers and thickening of others,—just such an appearance as might be expected to result from the occasional drifting of large beds of snow before they have time to become consolidated." *

The comparative absence of stones, gravel, or earth on the southern icebergs shows, likewise, the flat nature of the Antarctic ice-covering. "We certainly never saw," says Sir Wyville, "any trace of gravel or stones, or any foreign matter, necessarily derived from land, on an iceberg."

But supposing we should make the extravagant assumption that in this comparatively flat and uniform sheet the pressure, by some unexplained means is not evenly distributed, but that, on the contrary, it is all brought to bear on certain points and consumed in melting the ice, and that the total quantity of ice melted is the exact equivalent of the work performed by gravity; and let us further assume that the entire mass of the ice is already at the melting-point, and that, therefore, no work is required to raise its temperature, then the total quantity of ice melted would be of course $\frac{\cdot 9}{142}$, or $\frac{1}{158}$ of the entire mass. Gravity could perform only $\frac{1}{158}$ of the amount of work required to melt the entire sheet. If we suppose the sheet to be 1400 feet thick, then $\frac{1}{158}$ of this thickness will be equal to 9 feet. A layer of ice about 9 feet in thickness, therefore, is the total amount that gravity

* "Antarctic Regions," p. 16.

could, in such a case, *under any circumstances* have melted.

But more than this, it must be borne in mind that these 9 feet represent the total quantity which could be melted during the whole time the sheet was being formed; that is, from the time the bottom layer fell in the form of snow on the surface down to the present day. We have no means of ascertaining the length of this period. If we assume it to be 10,000 years, and this is probably an under-estimate, then 9 feet of ice melted during that period would amount to only 1 inch in 92 years, or $\frac{1}{92}$ of an inch annually. But whether the period be 10,000 or 5000 years, the quantity is so trifling that it may be practically disregarded in the present inquiry.

Nor is this all; for if the great mass of the ice be as much as 2° below the freezing-point, which it undoubtedly is, the total amount of heat generated by compression and friction during the 10,000 years would not suffice to raise the temperature of the ice even to the melting-point.

The Great Diminution in the Thickness of the Ice-strata from the top downwards not due, as supposed, either to Compression or to Melting.— The thinness of the lower as compared with the more superficial strata of the ice-sheet is considered by Sir Wyville Thomson to be mainly, if not altogether, due to two causes,—compression and melting of the ice, particularly the latter. "The regularity of this diminution," he says, "leaves it almost without a doubt that the layers observed are in the same category, and that therefore the diminution is due to subsequent pressure or other action upon a series of beds which were at the time of their deposition pretty nearly equally thick. About 60 or 80 feet from the top of an iceberg the strata of ice are a foot or so in thickness, although of a white colour, and thus indicating that they contain a quantity of air, and that the particles of ice are not in close apposition, are still very hard, and the specific gravity of the ice is not very much lower than that of layers not more than three inches thick nearer the water-line of the berg. Now it seems to me that this reduction cannot be due to compression alone, and that a portion of the substance of these lower layers must have been removed." *

If the layers three inches thick near the water-line were once a foot in thickness, as no doubt they were, then this great diminution in thickness cannot have been due to compression; for, had it been so, the density of those layers would be more than double that of water. But Sir Wyville has found that the specific gravity of the layers three inches thick is not much lower than of those a foot in thickness, which proves, as he has pointed out, that compression cannot account for their thinness; but it does not, as we shall presently see, necessarily prove "that a portion of the substance of these lower layers must have been removed."

Assuming that the lower layers were all originally of the same thickness as the upper, it has nevertheless been shown in Chapter V. that the gradual diminution in thickness of the layers from the top downwards follows independently altogether of compression or of the removal of any portion of the substance of the layers, either by melting or by any other means.

Ice radiating from a Centre of Dispersion becomes thinner, because the space over which it is spread

* "Antarctic Regions," p. 23.

becomes greater.—There is this peculiarity in continental ice, that there is a centre of dispersion from which the ice radiates in all directions. This is particularly true in reference to the Antarctic ice-cap. It does not necessarily follow that the centre of dispersion is the centre of the sheet. In the case of the Antarctic sheet the centre of dispersion cannot, however, be far from the Pole, and the Pole in all probability is not far from the centre of the sheet. We may therefore, in our inquiry, safely assume the Pole to be the centre of dispersion. It is obvious that, if the Antarctic ice be radiating in all directions from the Pole as a centre, a portion of a layer which in, say, latitude 85°, as was shown in Chapter V., covers 1 square foot of surface will, on reaching latitude 80°, cover 2 square feet. At latitude 70° it will occupy 4 square feet, and at latitude 60° the space covered will be 6 square feet. Then if the layer was 1 foot thick at latitude 85°, it would be only 6 inches thick at latitude 80°, 3 inches thick at latitude 70°, and 2 inches at latitude 60°. Had the square foot of ice come from latitude 89°, it would occupy 30 square feet by the time it reached latitude 60°, and its thickness would be reduced to 1-30th of a foot, or 2-5ths of an inch.

Now, the lower the layer the older it is, and the greater the distance which it has travelled. A layer near the bottom may have been travelling from the Pole for the past 10,000 or 15,000 years, whereas a layer near the top may perhaps not be twenty years old, and may not have travelled the distance of a mile. The ice at the bottom of a berg may have come from near the Pole, whereas the ice at the top may not have travelled 100 yards. It follows therefore that, other things being equal, the lower a layer is the thinner it should be, and that this is perfectly sufficient, as has already been stated, to account for the decrease in the thickness of the layers from the top downwards, without assuming any of the ice to have been removed by melting or by any other means.

Continental Ice radiating from a Centre of Dispersion must be Thickest at the Centre, and gradually Diminish in Thickness towards the Circumference.— Whatever theory we may adopt as to the cause of the motion of ice, it will follow as a necessary consequence that the sheet must be thickest at the centre and thinnest at its edge. In a continental sheet like that covering the Antarctic regions, we are not warranted, as has already been noticed, in assuming that the surface of the ground under the sheet slopes persistently outwards from the centre or Pole to the edge; in other words, we cannot infer that the Antarctic ice, like an ordinary glacier, rests on an inclined plane.

Now, if we adopt the generally accepted theory, that gravity is the force impelling the ice forward, we must assume the sheet to be thickest at the centre; for unless it were so, gravity could have no tendency to produce motion, because the force which moves the ice must not only act horizontally, but act more in one direction than in another; and this it could not do were the ice of uniform thickness. Were the sheet of this uniform thickness, the forces acting on it would balance each other, and no motion could result. If the sheet is to be forced out horizontally along the flat surface by *its own weight,* then there must be a piling up of the ice in the interior. If the ice comes from the centre, then the pressure must be greatest there; but in order to this, the sheet, of course, must be thickest at the centre.

Supposing it should be asserted that it is not the

pressure of the particle *a* that moves the particle *b* in front of it, and the pressure of the particle *b* that moves the particle *c*, and so on, but that each particle moves by its own weight, we are nevertheless led to the same conclusion. The weight of the particle (the force of gravity) will not move the particle unless the particle is allowed to descend. If a particle moves by its own weight from the centre of the sheet to the circumference, it must *descend:* it must pass from a *higher* to a *lower level.* It must move down an inclined plane from the centre to the circumference, but to allow it to do so the sheet must be thickest at the centre.*

If, on the other hand, we adopt the "Molecular" theory, or the "Dilatation" theory of the motion of the ice, or any other theory whatever which attributes the motion of the ice not to gravity, but to some expansive force acting in the interior of the mass, we are equally led to the same conclusion as to the greater thickness of the sheet at the centre. Although such a force will, of course, tend to push the ice as powerfully inwards in the direction of the Pole, or centre, as outwards in the direction of the circumference, yet the motion of the ice will always take place in the latter direction, and never in the former, for the latter will always be the direction of least resistance. The tendency of such a force is to produce an outward motion of the ice on

* That the entire mass of the Antarctic ice down to the bottom is in a state of motion, and not simply the upper layers, as some suppose, is demonstrable from the fact that icebergs are stratified down to their base. The iceberg is simply a piece broken off the edge of the sheet, and the stratified face of the berg is the counterpart of the edge from which it broke off; and as the icebergs are known to be stratified to their base, it proves that the sheet from which they were derived is likewise stratified to the bottom. The fact, therefore, that stratified icebergs are continually breaking off the Antarctic sheet, and have been for ages, proves that the sheet down to its bottom must have been in a state of outward motion.

the outer side, but to hold back or prevent such a motion taking place in the ice on the inner or poleward side. As such an expansive force is assumed to act in every portion of the mass, it follows that the nearer the outside of the sheet the more rapidly will the ice move, and consequently the thinner will the sheet become.

The Greater Thickness of the Sheet at the Pole independent of the amount of Snowfall at that place.—It has been proved that unless the Antarctic ice were thickest at the Pole and thinnest at the edge, motion could not take place. It follows, therefore, that however much the snowfall at the edge and other places may exceed that at the Pole, or centre of dispersion, the ice must always be thickest at the centre. For however small may be the snowfall, and consequent amount of ice formed annually at the Pole, snow and ice must of necessity continue to accumulate year by year till the sheet becomes thickest there. The ice at the Pole could not move out of its position till this were the case. Supposing there were no snow whatever falling at the Pole, and no ice being formed there, still the sheet would be thicker there than at the edge. For in this case the ice forming at some distance from the Pole all around would flow back, as has already been shown in Chapter V., towards the centre, and continue to accumulate there till the resistance to the inward flow became greater than the resistance to the outward; but this state would not be reached till the ice became thickest on the poleward side.

We have no reason to believe, however, that the quantity of snow falling at the Pole is not great. "One thing we know," says Sir Wyville Thomson, "that the precipitation throughout the Antarctic area is very great, and that it is always in the form of snow." Lieut. Wilkes, of the American Exploring Expedition, estimated the snowfall to be 30 feet per annum, and Sir James Ross says that during a whole month they had only three days free from snow. The very fact that perpetual snow is found at the sea-level at lat. 64° S. proves that the amount of precipitation in the form of snow in those regions must be great.

But there is one circumstance which must tend to make the snowfall near the Pole great, and that is the inflow of moist winds in all directions towards it; and as the area on which these currents deposit their snow becomes less and less as the Pole is reached, this must to a corresponding extent, as was shown in Chapter V., increase the quantity of snow falling on a given area.

Rate of Motion of the Antarctic Ice.—If we knew the rate at which the edge of the Antarctic ice-cap is advancing outwards, we could form a rough estimate of the amount of snowfall on the continent. Or, conversely, knowing the amount of snowfall, we could tell approximately the rate at which the ice is moving outwards.

Dr. Rink calculates that the yearly precipitation on Greenland in the form of snow and rain amounts to about 12 inches. About 2 inches he considers is carried off by ice into the sea, and the remaining 10 inches is carried to the sea in the form of sub-glacial rivers. He believes that the quantity disposed of by evaporation is trifling.

The amount of precipitation on the Antarctic continent is probably much greater than on Greenland. On the Antarctic continent it is all in the form of snow or hoar-frost, whereas in Greenland a considerable portion of it—in summer at least—is in the form of rain. For reasons already stated the proportionate amount carried off the Antarctic continent in the form of water to that of ice must be much less than on Greenland. The quantity of ice melted in the Antarctic regions from all causes, we have seen, cannot be great; and of that quantity the greater part must be re-solidified long before it can reach the sea. I can hardly think that it will be regarded as an over-estimate to affirm that at least one-half the precipitation must reach the sea in the form of ice. Assuming the annual precipitation to be no greater than that of Greenland, viz., 1 foot per annum, the quantity carried off in the form of ice would in this case be 6 inches. At what rate, then, would the edge of the cap require to be advancing outwards in order to discharge this 6 inches of ice? If we assume the cap to extend on an average down to latitude 70°, its area will be about 5,940,000 square miles, or 165,611,000,000,000 square feet. A layer 6 inches thick covering that area would contain 82,805,500,000,000 cubic feet of ice. The circumference of the cap is 45,300,000 feet, and its thickness at the edge is assumed, of course, to be 1400 feet. Were the ice, therefore, to move outwards at the rate of 1300 feet per annum, and to break up into bergs as it advanced, the quantity of ice discharged annually in the form of icebergs would be 82,446,000,000,000 cubic feet, an amount equal to the layer of ice 6 inches in thickness covering the area. Consequently, if 6 inches of ice be carried annually off the Antarctic continent, the edge of the cap must be moving outwards at the rate of about a quarter of a mile annually. Even supposing there were only 2 inches of ice discharged, the rate of motion would require to be between 400 and 500 feet per annum.

A quarter of a mile per annum cannot be regarded as an improbable rate of motion for continental ice, when we reflect that the Greenland ice has in some places a velocity ten times greater. Mr. Amund Helland, for example, found that the glacier of Jacobshaven has a velocity of about 20 metres *per diem*, which is upwards of 4 miles annually. The exceptional high velocity of the Greenland glaciers is no doubt owing to the fact that the ice-sheet covering that continent has to force its way through comparatively narrow outlets. If the sheet moved off the land in one unbroken mass, like the Antarctic sheet, its rate of motion would be much less.

It is the immense extent of the Antarctic continent which demands such a high velocity to get rid of the ice. To enable it to discharge the annual amount of ice, either the sheet must be excessively thick or its rate of motion excessively great. If, for example, the ice were only 700 feet instead of 1400 feet thick, its motion would require to be half a mile annually in order that the 6 inches of ice should be got rid of; while, if it were only 100 feet in thickness the rate of motion would need to be 3½ miles per annum.

It is this difficulty in getting away which is the chief cause of the enormous accumulation of ice on the Antarctic continent. And it is just this great thickness in the interior that enables the sheet to get rid of its superabundant ice. This is effected in two ways:—1st. The greater the thickness of the ice in the interior, the greater is the force by which it is impelled outwards, and, other things being equal, the greater is the velocity of the ice. 2nd. The thicker the sheet becomes, the greater is the quantity discharged corresponding to a given velocity. The velocity being the same, the quantity discharged is in proportion to the thickness of the sheet.

With the present rate of snowfall on the Antarctic continent it is physically impossible that the ice can be otherwise than of great thickness. Were not the sheet enormously thick the quantity of ice annually discharged would not equal that being formed, and consequently the ice would of necessity increase in thickness year by year, till the rate of discharge became equal to that of growth. We have just seen that it would require a thickness of not less than 1400 feet at the very edge of the cap to make the two rates equal, even although the ice was moving outwards with a velocity of a quarter of a mile per annum, a rate of motion greater than that of an Alpine glacier; and, on the other hand, to produce such a rate of motion as this, a thickness in the interior enormously greater than 1400 feet is required. If, from an increase in the snowfall, or from a decrease in the quantity of snow and ice melted, or from both combined, the annual amount of ice requiring to be discharged were doubled, the velocity remaining the same, the thickness of the sheet would ultimately become doubled also. Or, if the thickness of the sheet remained the same the velocity would be doubled. The actual result in such a case, however, would be that a restoration of equilibrium between supply and discharge would take place, by an increase both of thickness and velocity. As the quarter of a mile per annum of velocity would only be sufficient to discharge one-half the amount of ice being formed, the sheet would increase in thickness year by year. But this increase in thickness would produce an increase of velocity, and the increase both in thickness and velocity of motion would continue till the quantity of ice discharged would be equal to the 12 inches over the whole area, instead of the 6 inches as before. Equilibrium being now established, no further increase would take place either in the thickness of the sheet or in the velocity of its motion. If, on the contrary, the amount of ice being formed on the Antarctic continent were to become less than at present, both the thickness of the sheet and the velocity of its motion would become less.

The following conclusions have now been established:—

1. The Antarctic ice-sheet must be thickest at the centre of dispersion and thinnest at the edge.
2. The rate of motion of the ice must be least at the centre of dispersion and greatest at the edge.
3. The mean thickness of the edge of the sheet, other things being equal, must be proportional to the area of the sheet, and inversely as the rate at which the edge is moving outwards.
4. The area of the sheet, the thickness of its edge, the velocity of its motion outwards, the amount of snowfall, and the temperature of the regions are so related to one another that the value of any one of them can be determined approximately in terms of the others.

The Probable Thickness of the Ice at the Pole.—The point which now remains to be determined is, What is the thickness of the ice at the Pole, or centre of dispersion? The thickness of the sheet at the edge is admitted to be about 1400 feet, and this, as has been demonstrated, must be the thinnest part of the sheet. It must gradually thicken inwards towards the Pole as centre of dispersion, where the thickness reaches a maximum. How much thicker, then, must the sheet be at the centre than it is at the circumference? The question to be determined, stated in another form, is, What is the thickness of ice at the Pole required in order to impel the cap outwards in all directions at the rate of a quarter of a mile per annum, or even half that rate per annum? The upper surface of the sheet must slope upwards towards the centre or pole. What is the amount of this slope?

The Antarctic continent is generally believed to extend on an average from the South Pole down to about lat. 70° or so. In round numbers we may take the diameter of the continent at 2800 miles. The distance from the edge of the ice-cap to its centre, the Pole, will therefore be 1400 miles. A slope of 1 degree, continued for 1400 miles, will give 24 miles as the thickness of the ice at the Pole. But would a slope of 1 degree be sufficient to produce the required amount of motion? If the generally accepted theory of the cause of glacier motion be correct, it certainly would not. But supposing we assume that one-half or even one-quarter that amount of slope would suffice, still we have 6 miles as the thickness of the cap at the Pole.

To those who have not been accustomed to reflect on the physical conditions of this problem, this estimate may doubtless be regarded as somewhat extravagant; but a slight consideration will show that it would be even more extravagant to assume that a slope of less than half a degree would be sufficient to produce the necessary outflow of the ice. In estimating the thickness of a sheet of continental ice of one or two thousand miles across, our imagination is apt to deceive us. We can easily form a pretty accurate sensuous representation of the thickness of the sheet, but we can picture to ourselves no adequate representation of its superficial area. We can realise with tolerable accuracy a thickness of a few miles, but we cannot do this in reference to a superficial area 2800 miles across. Consequently, in judging what proportion the thickness of the sheet should bear to the superficial area, we are apt to fall into the error of under-estimating the thickness. We have a striking example of this in regard to the ocean. That which impresses us most forcibly in regard to the ocean is its profound depth. A mean depth of, say, 3 miles produces a striking impression; but if we could represent to the mind the vast area of the ocean as correctly as we can do its depth, *shallowness* rather than *depth* would be the impression produced. A sheet of water 100 yards in diameter, and only 1 inch deep, would not be called a *deep* but a very *shallow* pool or thin layer of water. But such a layer would be a corrcet representation of the ocean in miniature. Were we, in like manner, to represent to the eye in miniature the Antarctic ice-cap, even as 12 miles in thickness at the Pole, we should call it a *thin crust of ice.* The mean thickness of the sheet would be about 4 miles, and this would be represented by a carpet covering the floor of an ordinary-sized dining-room. Were those who consider the above estimate of the Antarctic ice-cap as extravagantly great called upon to sketch on paper a section of what they should deem a cap of moderate thickness, ninety-nine out of a hundred would draw one of much greater thickness than 12 miles at the centre.

The accompanying diagram represents a section across the cap drawn to a natural scale, the upper surface of the sheet having a slope of half a degree. No one looking at the section would pronounce it to be

SECTION ACROSS ANTARCTIC ICE-CAP, DRAWN TO A NATURAL SCALE.

Length represented by section = 2800 miles. Thickness of centre (South Pole) = 12 miles.
Slope of upper surface = half degree

too thick at the centre unless he were previously made aware that it represented a thickness of 12 miles at that place. The section, of course, is not intended to represent the actual thickness of the sheet, but to show how liable we are to over-estimate a thickness proportionate to an area so immense. It may here be mentioned that had the section been drawn upon a much larger scale—had it, for instance, been made 7 feet long instead of 7 inches—it would have shown to the eye in a more striking manner the thinness of the cap.

At the close of the reading of Prof. James Geikie's paper "On the Glacial Phenomena of the Long Island," before the Geological Society, in May, 1878, His Grace the Duke of Argyll stated that he doubted whether ice could move on a slope of 1 in 211. But a slope so small as 1 in 211 would give a thickness of seven miles at the Pole. Consequently, we have no alternative but to admit that a slope of 1 in 211 is sufficient, or the cap must be over seven miles thick at the Pole.*

* Prof. J. Geikie writes me as follows:—"I have given the height of the glaciation in the North-west Highlands as 3000 feet or thereabout, which taken in connection with the glacial phenomena of the Outer Hebrides, implies a slope for the surface of the ice-sheet of 1 in 211, or about 25 feet in the mile. It is not improbable, however, that a more detailed examination of the mainlands may compel us to admit a still greater thickness for the ice-sheet of the North-west—the surface of which may have reached to a height of 3500 feet in Ross-shire. This would yield a slope of 35 instead of 25 feet in the mile. After my paper had gone to press I received, through the kindness of Mr. George H. Cook, State Geologist of New Jersey, a copy of his Annual Report for 1877, in which the slope of the ice-sheet that flowed into the northern part of that State is estimated at 34 feet in the mile. Prof. Dana, you will remember, comes to the conclusion that the surface of the ice-sheet attained a height upon the Canadian water-shed of 12,000 feet, on the supposition that the ice sloped southwards at the rate of 10 feet in the mile,—if the slope were greater, the Canadian ice, of course, must have been thicker. The inclination of the ice-sheet in the area of the North Sea I estimate at about 12 or 13 feet in the mile."

Professor Nordenskjöld, as we have seen, found that the upper surface of the icy plain of Greenland has an elevation of 7000 feet, 280 miles from the coast. If the Antarctic ice-sheet has an equal slope, this would give 35,000 feet as the thickness of the ice at the Pole.

But to avoid all objections on the score of over-estimating the thickness of the cap, let us assume that a slope of an eighth of a degree, or less than one-half that of the Greenland sheet, would be sufficient to produce the necessary motion; the thickness of the sheet would of course be one-fourth that represented in the diagram, *but still it would be three miles thick at the Pole!*

There is another cause which tends to mislead us in forming an estimate of the actual thickness of the Antarctic ice. It is not in consequence of any *à priori* reason that can be urged against the probability of such a thickness of ice, but rather because it so far transcends our previous experience that we are so reluctant to admit such an estimate. If we never had any experience of ice thicker than what is found in England, we should feel startled on learning for the first time that, in the valleys of Switzerland, the ice lay from 200 to 300 feet in depth. Again, if we had never heard of glaciers thicker than those of Switzerland, we could hardly credit the statement that, in Greenland, they are actually from 2000 to 3000 feet thick. We in this country have long been familiar with Greenland; but till very lately no one ever entertained the idea that that continent was buried under one continuous mass of ice, with scarcely a mountain top rising above the icy mantle. And had it not been that the geological phenomena of the Glacial Epoch have for so many years accustomed our minds to such an extraordinary condition of things, Dr. Rink's description of the Greenland ice would probably have been regarded as the extravagant picture of a wild imagination.

The Ice of the Glacial Epoch.—The same general principles which we have been considering hold equally true in reference to the ice of the Glacial Epoch. Misapprehensions regarding the magnitude of continental ice lie at the very root of the opposition with which the Land-ice Theory of the chief phenomena of the Glacial Epoch has had to contend. One of the main objections urged against that theory is the magnitude of the ice-sheet which it demands. For example, to explain the glacial phenomena by the theory of land ice, we are compelled to infer that the whole of Scotland, Scandinavia, and the greater part of North-western Europe, were not only covered with ice, but covered to a depth of one or two thousand feet. But not only are the mainlands glaciated, but the islands of the Baltic, the Orkneys, the Shetlands, and the Hebrides, bear equal evidence of ice having passed over them. To explain this by the theory, we have further to assume that the ice-sheet which covered the land must have filled the Baltic, the German Ocean, and the surrounding seas; in short, that all these regions were buried underneath one continuous mass of ice.

To one with inadequate conceptions of the nature of continental ice, such a condition of things as this may appear incredible; but if the principles we have been considering be correct, it follows as a necessary consequence. If, during the Glacial Epoch, the quantity of ice annually formed in North-western Europe was much in excess of the quantity melted, enormous ice-sheets must of necessity have been formed.

The thickness of the sheet or sheets covering that region would depend, as has been shown, upon the area covered and the rate of snowfall, or, rather, the rate at which the ice was being formed. The sheet, as has also been shown, must have been thickest at the centre or centres of dispersion — if there were more than one — and thinnest at the edge. The extent of area covered by ice on North-western Europe must have been great; so also must have been the amount of snowfall.

That such a condition as this, to which we are led by theoretical considerations, did actually prevail during the Glacial Epoch is now established by the facts of observation. Norway we know was the great centre of dispersion of the ice, and here it has been found that the sheet attained its greatest thickness. It has been shown by Mr. Amund Helland that its thickness there was over a mile. Scotland was also a subordinate centre of dispersion, and we know that the ice moving off it was sufficient to prevent that country from being overridden by the great mass of ice flowing outwards in all directions from the Scandinavian centre. It was sufficient, but little more; for the Scandinavian ice, filling the German Ocean and passing over the Orkney and Shetland Islands, was so powerful as to bend back the Scottish ice and force it to turn round after it had entered the German Ocean, and pass obliquely over the flat lands of Caithness. It was also sufficient to fill the entire Baltic and to pass over on Germany, down even to the foot of the Saxon uplands. All this has now been completely established by the observations of geologists.

CHAPTER XV.

REGELATION AS A CAUSE OF GLACIER MOTION.

Why the Problem of Glacier Motion is so difficult.—Heat in Relation to Glacier Motion.—Regelation as a Cause of Motion.—Theories of the Cause of Regelation.—How Regelation produces Motion.—Heat transformed into Glacier Motion.

THE conditions which make the question of the cause of the descent of glaciers so perplexing seems to be this:—The ice of a glacier is not in a soft and plastic state, but is solid, hard, brittle, and unyielding. It nevertheless behaves in some respects in a manner very like what a soft and plastic substance would do if placed in similar circumstances, inasmuch as it accommodates itself to all the inequalities of the channel in which it moves. The ice of the glacier, though hard and solid, moves with a differential motion; the particles of the ice are displaced over each other, or, in other words, the ice shears as it descends. It had been concluded that the mere weight of the glacier is sufficient to shear the ice. Canon Moseley several years ago investigated this point,* and showed that it is not. He found that for a glacier to shear in the way that it is supposed to do, it would require a force some 30 or 40 times greater than the weight of the glacier. Consequently, for the glacier to descend, a force in addition to that of gravitation is required. What, then, is this force? It is found that

† Memoir read before the Royal Society, Jan. 7, 1869.

the rate at which the glacier descends depends upon the amount of heat which it is receiving. This shows that the motion of the glacier is in some way or other dependent upon heat. But in what respect can heat be regarded as a cause of motion? Heat cannot be directly a cause of motion. Neither can heat produce motion or displacement of the particles by making the ice soft and plastic; for we know that the ice of a glacier is not soft and plastic, but hard and brittle. Its proper function will be seen when considering the bearing of Regelation on glacier motion.

Whether or not regelation can be regarded as a cause of glacier motion will depend on the view which we may adopt as to the physical cause of regelation itself. There are three theories which have been advanced to explain regelation.

According to Professor James Thomson's theory, pressure is the cause of regelation. Pressure applied to ice tends to lower the melting-point, and thus to produce liquefaction, but the water which results is colder than the ice, and refreezes the moment it is relieved from pressure. When two pieces of ice are pressed together, a melting takes place at the points in contact, resulting from the lowering of the melting-point; the water formed, re-freezing, joins the two pieces together.

The objection which has been urged against this theory is that regelation will take place under circumstances where it is difficult to conceive how pressure can be regarded as the cause. Two pieces of ice, for example, suspended by silken threads in an atmosphere above the melting-point, if but simply allowed to touch each other, will freeze together. Professor J. Thomson, however, attributes the freezing to the pressure resulting from the capillary attraction of the two moist surfaces in contact. But when we reflect that it requires the pressure of a mile of ice—135 tons on the square foot—to lower the melting-point one degree, it must be obvious that the lowering effect resulting from capillary attraction in the case under consideration must be infinitesimal indeed.

The following clear and concise account of Faraday's theory, I quote from Professor Tyndall's "Forms of Water:"—

"Faraday concluded that *in the interior* of any body, whether solid or liquid, where every particle is grasped, so to speak, by the surrounding particles, and grasps them in turn, the bond of cohesion is so strong as to require a higher temperature to change the state of aggregation than is necessary *at the surface.* At the surface of a piece of ice, for example, the molecules are free on one side from the control of other molecules; and they, therefore, yield to heat more readily than in the interior. The bubble of air or steam in overheated water also frees the molecules on one side; hence the ebullition consequent upon its introduction. Practically speaking, then, the point of liquefaction of the interior ice is higher than that of the superficial ice.

"When the surfaces of two pieces of ice, covered with a film of the water of liquefaction, are brought together, the covering film is transferred from the surface to the centre of the ice, where the point of liquefaction, as before shown, is higher than at the surface. The special solidifying power of ice upon water is now brought into play *on both sides of the film.* Under these circumstances, Faraday held that the film would congeal, and freeze the two surfaces together."*

* "The Forms of Water," p. 173.

The following is the theory which was advanced in 'Climate and Time' to account for regelation :—

The freezing-point of water and the melting-point of ice, as Professor Tyndall remarks, touch each other, as it were, at this temperature 32° F. At a hair's-breadth lower water freezes; at a hair's-breadth higher ice melts. Now, if we wish, for example, to freeze water already just about the freezing-point, or to melt a piece of ice just about the melting-point, we can do this either by a change of temperature or by a change of the melting point. But it will be always much easier to effect this by the former than by the latter means. Take the case already referred to, of the two pieces of ice suspended in an atmosphere above the melting-point. The pieces at their surfaces are in a melting condition, and are surrounded by a thin film of water just an infinitesimal degree above the freezing-point. The film has on the one side solid ice at the freezing-point, and on the other a warm atmosphere considerably above the freezing-point. The tendency of the ice is to lower the temperature of the film, while that of the air is to raise its temperature. When the two pieces are brought into contact the two films unite and form one film separating the two pieces of ice. This film is not like the former in contact with ice on the one side and warm air on the other. It is surrounded on both sides by solid ice. The tendency of the ice, of course, is to lower the film to the same temperature as the ice itself, and thus to produce solidification. It is evident that the film must either melt the ice or the ice must freeze the film, if the two are to assume the same temperature. But the power of the ice to produce solidification, owing to its greater mass, is enormously greater than the power of the film to produce fluidity, consequently regelation is the result.

Let us now consider the bearing which the foregoing principles have on glacier motion. Heat can pass through a mass of ice, either by *radiation* or by the process of *conduction,* without visibly destroying the solidity of the ice. This has been proved experimentally by Professor Tyndall.* Although the general solidity of the ice of a glacier is not sensibly affected by the passage of the heat, nevertheless a process of melting may be incessantly going on in the interior of the ice; and as the effects of melting would doubtless be counterbalanced by the opposite process of regelation, the general solidity of the ice would thus be maintained. That the passage of heat through ice will melt particles in the interior is no mere hypothesis, but a fact which has been established both by observation and by experiment. Owing to the principle of regelation a particle melted in the interior would, in all probability, in nine cases out of ten, re-solidify. Let us now consider what would probably be the behaviour of the melted particle under such conditions, and the bearing which its solidification would have on glacier motion.

Ice is evidently not absolutely solid throughout. It is composed of crystalline particles, which, though in contact with one another, are, however, not packed together so as to occupy the least possible space, and, even though they were, the particles would not fit so closely together as to exclude interstices. The crystalline particles are, however, united to one another at special points determined by their polarity, and on this account they require more space; and this in all probability is the reason why ice, volume for volume, is less dense than water. It is obvious that when a crystalline particle melts it will not merely tend to

* See "Heat as a Mode of Motion," Appendix to Chap. IX.

descend by its weight into any space which it may find, but capillary attraction will cause it to flow into interstices between adjoining crystalline particles; but owing to the principles of regelation, already discussed, the thin film of water would instantly become re-solidified. It would not, however, solidify so as to fit the cavity which it occupied when in the fluid state, for the liquid particle in solidifying assumes the crystalline form, and of course there will be a definite proportion between the length, breadth, and thickness of the crystal; consequently it will always happen that the interstice in which it solidifies will be too narrow to contain it. The result will be that the fluid particle, in passing into the crystalline form, will press the two adjoining particles aside in order to make sufficient room for itself between them, and this it will do, no matter what amount of space it may possess in all other directions. The crystal will not form to suit the cavity, the cavity must be made to contain the crystal. And what holds true of one particle, holds true of every particle which melts and re-solidifies. This process is no doubt going on incessantly in every part of the glacier, and in proportion to the amount of heat which the glacier is receiving. This internal pressure, resulting from the solidifying of the fluid particles in the interstices of the ice, acts on the mass of the ice as an expansive force, tending to cause the glacier to widen out laterally in all directions.

Conceive a mass of ice lying on a flat horizontal surface, and receiving heat on its upper surface, say from the sun; as the heat passes downwards through the mass, particles of the ice melt and re-solidify. Each fluid particle solidifies in an interstice, which has to be widened in order to contain it. The pressure thus exerted by the continual re-solidifying of the particles will cause the mass to widen out laterally, and of course as the mass widens out it will grow thinner and thinner if it does not receive fresh acquisition on its surface. In the case of a glacier lying in a valley, motion, however, will only take place in one direction. The sides of the valley prevent the glacier from widening; and as gravitation opposes the motion of the ice up, and favours its motion down the valley, the path of least resistance to the pressure produced by regelation will always be down the slope, and consequently in this direction displacement will take place. Molecular pressure will therefore produce motion in the same direction as that of gravity. In other words, it will tend to cause the glacier to descend the valley.

The lateral expansion of the ice from internal pressure explains in a clear and satisfactory manner how rock-basins may be excavated by means of land-ice. It also removes the difficulties which have been felt in accounting for the ascent of ice up a steep slope. The main difficulty besetting the theory of the excavation of rock-basins by ice is to explain how the ice after entering the basin manages to get out again—how the ice at the bottom is made to ascend the sloping sides of the basin. Pressure acting from behind, it has been argued by some, will simply cause the ice lying above the level of the basin to move forward over the surface of the mass filling it. This conclusion is, however, incorrect. The ice filling the basin and the glacier overlying it are united in one solid mass, so that the latter cannot move over the former without shearing; and although the resistance to motion offered by the sloping sides of the basin may be much greater than the resistance to shear, still the ice will be slowly dragged out of the basin. However, in order to obviate this objection to which I refer, the advocates of the glacial origin of lake-basins point out that the length of those basins in proportion to their depth is so great that the slope up which the ice has to pass is in reality but small. This, no doubt, is true of lake-basins in general, but it does not hold universally true. But the theory here advocated does not demand that an ice-formed lake-basin cannot have steep sides. We have incontestible evidence that ice will pass up a steep slope; and, if ice can pass up a steep slope, it can excavate a basin with a steep slope. That ice will ascend a steep slope is proved by the fact that comparatively deep and narrow river-valleys, such as that of the Tay in some places,* are found often striated across. Hills, also, which stood directly in the path of the ice of the Glacial Epoch are sometimes found striated *upwards* from their base to their summit.

From what has been already stated in reference to the re-solidifying of the particles in the interstices of the ice, the application of the theory to the explanation of the effects under consideration will no doubt be apparent. Take the case of the passage of the ice-sheet across a river-valley. As the upper surface of the ice-sheet is constantly receiving heat from the sun and the air in contact with it, there is consequently a transference of heat from above downwards to the bottom of the sheet. This transference of heat is accompanied by the melting and re-solidifying of successive particles in the manner already detailed. As the fluid particles tend to flow into adjoining interstices before solidifying and assuming the crystalline form, the interstices of the ice at the bottom of the valley are constantly being filled by fluid particles from above. These particles no sooner enter the interstices than they pass into the crystalline form, and

* See 'Climate and Time,' p. 526.

become, of course, separated from their neighbours by fresh interstices, which new interstices become filled by fluid particles which, in turn, crystallize, anew forming interstices, and so on. The ice at the bottom of the valley, so long as this process continues, is constantly receiving additions from above. The ice must therefore expand laterally to make room for these additions, which it must do unless the resistance to lateral expansion be greater than the force exerted by the fluid particles in crystallizing. But a resistance sufficient to do this must be enormous. The ice at the bottom of the valley cannot expand laterally without passing up the sloping sides. In expanding it will take the path of least resistance, but the path of least resistance will always be on the side of the valley towards which the general mass of the ice above is flowing.

We can from these conditions understand how the softer portions of the rocky surface over which the ice moved should have been excavated into hollow basins. We have also an explanation of the transport of boulders from a lower to a higher level, for if ice can move from a lower to a higher level, it of course can carry boulders along with it.

Heat Transformed into Glacial Motion.—From what has been stated regarding the cause of glacial motion, it will now be obvious that a considerable portion of the sun's heat entering the ice must be transformed into work in the motion of the glacier. When a particle of ice is melted and then re-solidified, the amount of heat evolved during solidification is equal to that which had been expended in melting the particle. The particle in solidifying expands, and if, in expanding, work is performed by the expanding particle, the amount of heat evolved during solidification will then fall short of that which had been expended in melting by an amount which is exactly the equivalent of the work performed. The equation will be thus:—

Heat expended in melting = Heat evolved in re-freezing + Work performed.

A portion of the heat expended in melting is thus transformed into work. Now, if the motion of a glacier be mainly due to the expansive force exerted by the melted particles of the ice during their solidification, then the original source of this motion must be the heat received from the sun. Glacial motion must, therefore, in so far as it is the result of regelation, be transformed heat, or, in other words, molecular motion transformed into glacier motion.

CHAPTER XVI.

THE TEMPERATURE OF SPACE AND ITS BEARING ON TERRESTRIAL PHYSICS.

The Importance of Knowing the Temperature of Space. — The Researches of Pouillet and Herschel in reference to the Temperature of Space.—A Defect in Dulong's and Petit's Formula. — Professor Balfour Stewart on Radiation of Thin Plates. — Radiation of Gases.

FEW questions bearing directly on terrestrial physics have been so much overlooked as that of the temperature of stellar space; that is to say, the temperature which a thermometer would indicate if placed at the outer limits of our atmosphere and exposed to no other influence than that of radiation from the stars. Were we asked what was probably the mid-winter temperature of our island 11,700 years ago, when the winter solstice was in aphelion, we could not tell unless we knew the temperature of space. Again, without a knowledge of the temperature of space, it could not be ascertained how much the temperature of the North Atlantic and the air over it were affected by the Gulf Stream. We can determine the quantity of heat conveyed into the Atlantic by the stream, and compare it with the amount received by that area directly from the sun, but this alone does not enable us to say how much the temperature is raised by the heat conveyed. We know that the basin of the North Atlantic receives from the Gulf Stream a quantity of heat equal to about one-fourth that received from the sun, but unless we know the temperature of space we cannot say how much this one-fourth raises the temperature of the Atlantic. Suppose 56° to be the temperature of that ocean; this is 517° of absolute temperature which is derived from three sources, viz.: (1) direct heat from the sun, (2) heat from the Gulf Stream, and (3) heat from the stars. Now, unless we know what proportion the heat of the stars bears to that of the sun, we have no means of knowing how much of the 517° is due to the stars, and how much to the sun or to the Gulf Stream.

M. Pouillet, Sir John Herschel, and Professor Langley are the only physicists who appear to have devoted attention to the problem. M. Pouillet came to the conclusion that space has a temperature of – 142° C. or – 224° F., and Sir John Herschel, following a different method of inquiry, arrived at nearly the same result, viz., that its temperature is about – 239° F.

Can space, however, really have so high a temperature as – 239°? Absolute zero is – 461°. Space in this case would have an absolute temperature of 222°, and consequently our globe would be nearly as much indebted to the stars as to the sun for its heat. If so, space must be enormously more transparent to heat rays than to light rays. If the heat of the stars be as feeble as their light, space cannot be much above absolute zero, and this is the opinion expressed to me a short time ago by one of the most eminent physicists of the day. Professor Langley is also of this opinion; for he concludes that the amount of heat received from the sun is to that derived from space as much as four to one; and consequently if our luminary were extinguished, the temperature of our earth would fall to about – 360° F.

It must be borne in mind that Pouillet's Memoir

was written more than forty years ago, when the data available for elucidating the subject were far more imperfect than now, especially as regards the influence of the atmosphere on radiant heat. For example, Pouillet comes to the conclusion that, owing to the fact of our atmosphere being less diathermanous to radiation from the earth than to radiation from the sun and the stars, were the sun extinguished the radiation of the stars would still maintain the surface of our globe at $-89°$ C., or about $53°$ C. above that of space. The experiments of Tyndall, however, show that the absorbing power of the atmosphere for heat-rays is due almost exclusively to the small quantity of aqueous vapour which it contains. It is evident, therefore, that but for the sun there would probably be no aqueous vapour, and consequently nothing to protect the earth from losing its heat by radiation. Deprived of solar heat, the surface of the ground would sink to about as low a temperature as that of stellar space, whatever that temperature may actually be.

But before we are able to answer the foregoing questions, and tell, for example, how much a given increase or decrease in the *quantity* of sun's heat will raise or lower the *temperature*, there is another physical point to be determined, on which a considerable amount of uncertainty still exists. We must know in what way the temperature varies with the amount of heat received. In computing, say, the rise of temperature resulting from a great increase in the quantity of heat received, should we assume with Newton that it is proportional to the increase in the quantity of heat received, or should we adopt Dulong's and Petit's formula?

In estimating the extent to which the temperature of the air would be affected by a change in the sun's distance, I have hitherto adopted the former mode. This probably makes the change of temperature too great, while Dulong's and Petit's formula, adopted by Mr. Hill ("Nature," vol. xx. p. 626), on the other hand, makes it too small. Dulong's and Petit's formula is an empirical one, which has been found to agree pretty closely with observation within ordinary limits, but we have no reason to assume that it will hold equally correct when applied to that of space, any more than we have to infer that it will do so in reference to temperature as high as that of the sun. When applied to determine the temperature of the sun from his rate of radiation, it completely breaks down, for it is found to give only a temperature of 2130° F. ("Amer. Jour. Science," July, 1870), or not much above that of an ordinary furnace.

But besides all this it is doubtful if it will hold true in the case of gases. From the experiments of Prof. Balfour Stewart ("Trans. Edin. Roy. Soc.", xxii.) on the radiation of glass plates of various thicknesses, it would seem to follow that the radiation of a material particle is probably proportionate to its absolute temperature, or, in other words, that it obeys Newton's law. Prof. Balfour Stewart found that the radiation of a thick plate of glass increases more rapidly than that of a thin plate as the temperature rises, and that, if we go on continally diminishing the thickness of the plate whose radiation at different temperatures we are ascertaining, we find that, as it grows thinner and thinner, the rate at which it radiates its heat as its temperature rises becomes less and less. In other words, as the plate grows thinner its rate of radiation becomes more and more proportionate to its absolute temperature. And we can hardly resist the conviction that if it were possible to go on diminishing the thickness of the plate till we reached a film so thin as to embrace but only one particle in its thickness, its rate of radiation would be proportionate to its temperature, or, in other words, it would obey Newton's law. Prof. Balfour Stewart's explanation is this: As all substances are more diathermanous for heat of high than of low temperatures, when a body is at a low temperature only the exterior particles supply the radiation, the heat from the interior particles being all stopped by the exterior ones, while at a high temperature part of the heat from the interior is allowed to pass, thereby swelling the total radiation. But as the plate becomes thinner and thinner, the obstructions to interior radiation become less and less, and as these obstructions are greater for radiation at low than high temperatures, it necessarily follows that, by reducing the thickness of the plate, we assist radiation at low more than at high temperatures.

If this be the true explanation why the radiation of bodies deviates from Newton's law, it should follow that in the case of gases where the particles stand at a considerable distance from one another, the obstruction to interior radiation must be far less than in a solid, and consequently that the rate at which a gas radiates its heat as its temperature rises, must increase more slowly than that of a solid substance. In other words, in the case of a gas, the rate of radiation must correspond more nearly to the absolute temperature than in that of a solid; and the less the density and volume of a gas, the more nearly will its rate of radiation agree with Newton's law. The obstruction to interior radiation into space must diminish as we ascend in the atmosphere, at the outer limits of which, where there is no obstruction, the rate of radiation should be pretty nearly proportional to the absolute temperature. May not this to a certain extent be the cause why the temperature of the air diminishes as we ascend?

If the foregoing considerations be correct, it ought to follow that a reduction in the amount of heat received from the sun, owing to an increase of his distance, should tend to produce a greater lowering effect on the temperature of the air than it does on the temperature of the solid ground. Taking, therefore, into consideration the fact that space has probably a lower temperature than − 239°, and that the temperature of our climate is determined by the temperature of the air, it will follow that the error of assuming that the decrease of temperature is proportional to the decrease in the intensity of the sun's heat may not be great.

In estimating the extent to which the winter temperature is lowered by a great increase in the sun's distance, there is another circumstance which must be taken into account. The lowering of the temperature tends to diminish the amount of aqueous vapour contained in the air, and this in turn tends to lower the temperature by allowing the air to throw off its heat more freely into space.

CHAPTER XVII.

THE PROBABLE ORIGIN AND AGE OF THE SUN'S HEAT.

Age of the Sun's Heat according to the Gravitation Theories.—Testimony of Geology as to the Age of Life on the Globe.—Evidence from "Faults."—Rate of Denudation.—Age of the Stratified Rocks as determined by the Rate of Denudation.

Age of the Sun's Heat according to the Gravitation Theories.—The total annual amount of radiation from the whole surface of the sun is 8340×10^{30} foot-pounds. To maintain the present rate of radiation it would require the combustion of about 1500 lbs. of coal per hour on every square foot of the sun's surface; and were the sun composed of that material it would all be consumed in less than 5000 years. The opinion that the sun's heat is maintained by combustion cannot be entertained for a single moment. Mr. Lockyer has suggested that the elements of the sun are, owing to its excessive temperature, in a state of dissociation, and some have supposed that this fact might help to explain the duration of the sun's heat. But it must be obvious that, even supposing we were to make the most extravagant estimate of the chemical affinities of these elements, the amount of heat derived from their combination could at most give us only a few thousand years additional heat. Under every conceivable supposition, the combustion theory must be abandoned.

It is now generally held by physicists that the

enormous store of heat possessed by the sun could only have been derived from gravitation. For example, a pound of coal falling into the sun from an infinite distance would produce by its concussion more than 6000 times the amount of heat that would be generated by its combustion. It would, in fact, amount to upwards of 65,000,000,000 foot-pounds—an amount of energy sufficient to raise 1000 tons to a height of $5\frac{1}{2}$ miles.

There are two forms in which the gravitation theory has been presented: the first, the meteoric theory, propounded by Dr. Meyer: and the second, the contraction theory, advocated by Helmholtz. The meteoric theory of the sun's heat has now been pretty generally abandoned for the contraction theory advanced by Helmholtz. Suppose, with Helmholtz, that the sun originally existed as a nebulous mass, filling the entire space presently occupied by the solar system, and extending into space indefinitely beyond the outermost planet. The total amount of work in foot-pounds performed by gravitation in the condensation of this mass to an orb of the sun's present size can be found by means of the following formula given by Helmholtz:—

$$\text{Work of condensation} = \frac{3}{5} \cdot \frac{r^2 M^2}{Rm} \cdot g \,.$$

M is the mass of the sun, m the mass of the earth, R the sun's radius, and r the earth's radius. Taking—

$M = 4230 \times 10^{27}$ lbs., $m = 11{,}920 \times 10^{21}$ lbs.

$R = 2{,}328{,}500{,}000$ feet, and $r = 20{,}889{,}272$ feet,

we have then, for the total amount of work performed by gravitation in foot-pounds,

$$\frac{3}{5} \cdot \frac{(20{,}889{,}272{\cdot}5)^2 \times (4230 \times 10^{27})^2}{2{,}328{,}500{,}000 \times 11{,}920 \times 10^{21}}$$

$$= 168{,}790 \times 10^{36} \text{ foot-pounds.}$$

The amount of heat thus produced by gravitation would suffice for 20,237,500 years.

The conclusions are based upon the assumption that the density of the sun is uniform throughout. But it is highly probable that the sun's density increases towards the centre, in which case the amount of work performed by gravitation would be something more than the above.

Testimony of Geology as to the Age of Life on the Globe.

At this point, in reference to the age of our globe, geology and physics are generally supposed to come into direct antagonism. For if it be true, as physicists maintain, that gravitation is the only possible source from which the sun could have derived its store of energy, then the sun could not have maintained our globe at its present temperature for more than about 20 millions of years. "On the very highest computation which can be permitted," says Professor Tait, "it cannot have supplied the earth, even at the present rate, for more than about fifteen or twenty million years." * The limit to the age of the sun's heat must have limited the age of the habitable globe. All the geological history of the globe would necessarily be comprehended within this period. If the sun derived its heat from the condensation of its mass, then it could not possibly be more than about twenty million years since the beginning of the Laurentian period. But twenty million years would be considered by most geologists to represent only a comparatively small portion of the time which must have elapsed since organic life began on our globe.

* "Recent Advances in Physical Science," p. 175.

It is true that the views which formerly prevailed amongst geologists, in regard to the almost unlimited extent of geological time, have of late undergone very considerable modifications; but there are few geologists, I presume, who would be willing to admit that the above period is sufficient to comprehend the entire history of stratified rocks.

It is the facts of denudation which most forcibly impress the mind with a sense of immense duration, and show most convincingly the great antiquity of the earth.

We know unquestionably that many of the greatest changes undergone by the earth's crust were produced, not by convulsions and cataclysms of nature, but by those ordinary agencies that we see at work every day around us, such as rain, snow, frost, ice, chemical action, &c. Valleys have not been produced by violent dislocations, nor the hills by upheavals, but both have been carved out of the solid rock by the silent and gentle agency of chemical action, frost, rain, ice, and running water. In short, the rocky face of our globe has been moulded into hill and dale, and ultimately worn down to the sea-level by means of these apparently trifling agents, not merely once or twice, but probably dozens of times over during past ages. Now, when we reflect that with such extreme slowness do these agents perform their work that we might, if we could, watch their operations from year to year, and from century to century, without being able to perceive that they make any sensible impression, we are necessitated to conclude that geological periods must be enormous. The utter inadequacy of a period of 20 million years for the age of our earth is demonstrable from the enormous thickness of rock which is known to have been removed off certain areas by denudation. I shall now briefly refer to a few of the many facts which might be adduced on this point.

Evidence from "Faults."—One plain and obvious method of showing the great extent to which the general surface of the country has been lowered by denudation is furnished, as is well known, by the way in which the inequalities of surface produced by faults or dislocations have been effaced. It is quite common to meet with faults where the strata on the one side have been depressed several hundreds—and in some cases thousands—of feet below that on the other, but we seldom find any indications of such on the surface, the inequalities on the surface having been all removed by denudation. But in order to effect this a mass of rock must have been removed equal in thickness to the extent of the dislocation. The following are a few examples of large faults:—

The great Irwell fault, described by Prof. Hull,* which stretches from the Mersey west of Stockport to the north of Bolton, has a throw of upwards of 3000 feet.

Some remarkable faults have been found by Prof. Ramsay in North Wales. For example, near Snowdon, and about a mile E.S.E. of Beddgelert, there is a fault with a downthrow of 5000 feet; and in the Berwyn Hills, between Bryn-mawr and Post-gwyn, there is one of 5000 feet. In the Aran Range there is a great fault, designated the Bala fault, with a downthrow of 7000 feet. Again, between Aran Mowddwy and Careg Aderyn the displacement of the strata amounts to no less than from 10,000 to 11,000 feet.† Here we have evidence that a mass of rock, varying from 1 mile to 2 miles in vertical thickness, must have been denuded in

* Mem. Geol. Survey of Lancashire, 1862.

† Mem. Geol. Survey of Great Britain, vol. iii.

many places from the surface of the country in North Wales.

The fault which passes along the east side of the Pentlands is estimated to have a throw of upwards of 3000 feet.* Along the flank of the Grampians a great fault runs from the North Sea at Stonehaven to the estuary of the Clyde, throwing the Old Red Sandstone on end sometimes for a distance of 2 miles from the line of dislocation. The amount of the displacement, Prof. Geikie† concludes, must be in some places not less than 5000 feet, as indicated by the position of occasional outlyers of conglomerate on the Highland side of the fault.

The great fault crossing Scotland from near Dunbar to the Ayrshire coast, and which separates the Silurians of the South of Scotland from the Old Red Sandstone and Carboniferous tracts of the North, has been found, by Mr. B. N. Peach, of the Geological Survey,‡ to have in some places a throw of fully 15,000 feet. This great dislocation is older than the Carboniferous period, as is shown by the entire absence of any Old Red Sandstone on the south side of the fault, and by the occurrence of the Carboniferous Limestone and Coal-measures lying directly on the Silurian rocks. We obtain here some idea of the enormous amount of denudation which must have taken place during a comparatively limited geological epoch. So vast a thickness of Old Red Sandstone could not, as Mr. Peach remarks, "have ended originally where the fault now is, but must have swept southwards over the Lower Silurian uplands. Yet these thousands of feet of sandstones, conglomerates, lavas, and tuffs were so

* Memoir to sheet 32, Geol. Survey Map of Scotland.

† "Nature," vol. xiii., p. 390.

‡ Explanation to Sheet 15, Geol. Survey Map of Scotland.

completely removed from the south side of the fault previous to the deposition of the Carboniferous Limestone series and the Coal-measures that not a fragment of them is anywhere to be seen between these latter formations and the old Silurian floor." This enormous thickness of nearly 3 miles of Old Red Sandstone must have been denuded away during the period which intervened between the deposition of the Lower Old Red Sandstone and the accumulation of the Carboniferous Limestone.

Near Tipperary, in the south of Ireland, there is a dislocation of the strata of not less than 4000 feet,* which brings down the coal-measures against the silurian rocks. Here 1000 feet of Old Red Sandstone, 3000 feet of Carboniferous Limestone, and 800 feet of Coal-measures have been removed by denudation off the Silurian rocks. Not only has this immense thickness of beds been carried away, but the Silurian itself on which they rested has been eaten down in some places into deep valleys several hundreds of feet below the surface on which the Old Red Sandstone rested.

Faults to a similar extent abound on the Continent and in America, but they have not been so minutely examined as in this country. In the valley of Thessolon, to the north of Lake Huron, there is a dislocation of the strata to the extent of 9000 feet.†

In front of the Chilowee Mountains there is a vertical displacement of the strata of more than 10,000 feet.‡ Prof. H. D. Rogers found in the Appalachian coal-fields faults ranging from 5000 feet to more than 10,000 feet of displacement.

There are other modes than the foregoing by means

* Jukes's and Geikie's "Manual of Geology," p. 441.

† "Geology of Canada, 1863," p. 61.

‡ Safford's "Geology of Tennessee," p. 309.

of which geologists are enabled to measure the thickness of strata which may have been removed in places off the present surface of the country, into the details of which I need not here enter. But I may give a few examples of the enormous extent to which the country, in some places, has been found to have been lowered by denudation.

Prof. Geikie has shown* that the Pentlands must at one time have been covered with upwards of a mile in thickness of Carboniferous rocks which have all been removed by denudation.

In the Bristol coal-fields, between the River Avon and the Mendips, Prof. Ramsay has shown† that about 9000 feet of Carboniferous strata have been removed by denudation from the present surface.

Between Bendrick rock and Garth Hill, South Glamorganshire, a mass of Carboniferous and Old Red Sandstone, of upwards of 9000 feet, has been removed. At the Vale of Towy, Caermarthenshire, about 6000 feet of Silurian and 5000 feet of Old Red Sandstone—in all about 11,000 vertical feet—have been swept away. Between Llandovery and Aberaeron a mass of about 12,000 vertical feet of the Silurian series has been removed by denudation. Between Ebwy and the Forest of Dean, a distance of upwards of 20 miles, a thickness of rock varying from 5000 to 10,000 feet has been abstracted.

Prof. Hull found‡ on the northern flanks of the Pendle Range, Lancashire, the Permian beds resting on the denuded edges of the Millstone Grit, and these were again observed resting on the Upper Coal-measures south of the Wigan coal-field. Now, from

* Mem. to Sheet 32, Geol. Survey of Scotland.

† "Denudation of South Wales." Memoirs of Geol. Survey, vol. i.

‡ "Quart. Journ. Geol. Soc.," vol. xxiv., p. 323.

the known thickness of the Carboniferous series in this part of Lancashire, he was enabled to calculate approximately the quantity of Carboniferous strata which must have been carried away between the period of the Millstone Grit and the deposition of the Permian beds, and found that it actually amounted to no less than 9,900 feet. He also found in the Vale of Clitheroe, and at the base of the Pendle Range, that the Coal-measures, the whole of the Millstone Grit, the Yoredale series, and part of the Carboniferous Limestone, amounting in all to nearly 20,000 feet, had been swept away—an amount of denudation which, as Prof. Hull remarks, cannot fail to impress us with some idea of the prodigious lapse of time necessary for its accomplishment.

In the Nova Scotia coal-fields one or two miles in thickness of strata have been removed in some places.*

It may be observed that, enormous as is the amount of denudation indicated by the foregoing figures, these figures do not represent in most cases the actual thickness of rock removed from the surface. We are necessitated to conclude that a mass of rock equal to the thickness stated must have been removed, but we are in most cases left in uncertainty as to the total thickness which has actually been carried away. In the case of a fault, for example, with a displacement of (say) one mile, where no indication of it is seen at the surface of the ground, we know that on one side of the fault a thickness of rock equal to one mile must have been denuded, but we do not know how much more than that may have been removed. For anything which we know to the contrary, hundreds of feet of rock may have been removed before the dislocation took place, and as many more hundreds

* Lyell's "Student's Manual," chap. 23.

after all indications of dislocation had been effaced at the surface.

But it must be observed that the total quantity of rock which has been removed from the *present* surface of the land is evidently small in proportion to the total quantity removed during the past history of our globe. For those thousands and thousands of feet of rock which have been denuded were formed out of the waste of previously existing rocks, just as these had been formed out of the waste of yet older rock-masses. In short, as a general rule, the rocks of one epoch have been formed out of those of preceding periods, and go themselves to form those of subsequent epochs.

In many of the cases of enormous denudation to which we have referred, the erosion has been effected during a limited geological epoch. We have, for example, seen that upwards of a mile in thickness of Carboniferous rock has been denuded in the area of the Pentlands. But the Pentlands themselves, it can be proved, existed as hills, in much their present form, before the Carboniferous rocks were laid down over them; and as they are of Lower Old Red Sandstone age, and have been formed by denudation, they must consequently have been carved out of the solid rock between the period of the Old Red Sandstone and the beginning of the Carboniferous age. This affords us some conception of the immense lapse of time represented by the Middle and Upper Old Red Sandstone periods.

Again, in the case of the great fault separating the Silurians of the south of Scotland from the Old Red Sandstone tracts lying to the north, a thickness of the latter strata of probably more than a mile, as we have seen, must have been removed from the ground to the south of the fault before the commencement of the

Carboniferous period. And again, in the case of the Lancashire coal-fields, to which reference has been made, nearly two miles in thickness of strata had been removed in the interval which elapsed between the Millstone Grit and the Permian periods.

Rate of Denudation.—As we are enabled, from geological evidence, to form some rough estimate of the extent to which the country in various places has been lowered by sub-aërial denudation during a given epoch, it is evident that we should have a means of arriving at some idea of the length of that epoch, did we know the probable rate at which the denudation took place. If we had a means of forming even the roughest estimate of the probable average rate of sub-aërial denudation during past ages, we should be enabled thereby to assign approximately an inferior limit to the age of the stratified rocks. We could then tell, at least, whether the amount of sub-aërial denudation known to have been effected during past geological ages could have been accomplished within 20 million years or not, and this is about all with which we are at present concerned. And if it can be proved that a period of 20 millions of years is much too short to account for the amount of denudation known to have taken place, then it is certain that the gravitation theory cannot explain the origin and source of the sun's heat.

A very simple and obvious method of determining the present mean rate of sub-aërial denudation was pointed out by me several years ago,* viz., that the rate of denudation must be equal to the rate at which the materials are carried off the land into the sea. But the rate at which the materials are thus abstracted

* "Phil. Mag.," May, 1868; Feb., 1867. 'Climate and Time,' Chapter XX.

is measured by the rate at which sediment is carried down by our rivers. Consequently, in order to determine the present rate of sub-aërial denudation, we have only to ascertain the quantity of sediment annually carried down by the river systems.

Very accurate measurements have been made of the quantity of sediment carried down into the Gulf of Mexico by the River Mississippi, and it is found to amount to 7,474,000,000 cubic feet per annum. The area drained by the river is 1,224,000 square miles. Now, 7,474,000,000 cubic feet removed from 1,224,000 square miles of surface is equal to 1-4566th of a foot off the surface per annum, or 1 foot in 4566 years. The specific gravity of the sediment is taken at 1·9, and that of the rock at 2·5; consequently the amount removed is equal to 1 foot of rock in about 6000 years. For many reasons there are few rivers better adapted for affording us a fair average of the rate of sub-aërial denudation than the Mississippi. In this connection I may here quote the words of Sir Charles Lyell:—"There seems," he says, "no danger of our over-rating the mean rate of waste by selecting the Mississippi as our example, for that river drains a country equal to more than half the continent of Europe, extends through 20 degrees of latitude, and therefore through regions enjoying a great variety of climate, and some of its tributaries descend from mountains of great height. The Mississippi is also more likely to afford us a fair test of ordinary denudation, because, unlike the St. Lawrence and its tributaries, there are no great lakes in which the fluviatile sediment is thrown down and arrested on its way to the sea." *

* "Student's Manual of Geology," p. 91 (second edition).

Rough estimates have been made of the sediment carried down by some eight or ten European rivers; and although those estimates cannot be depended upon as being anything like accurate, still they show that it is extremely probable that the European continent is being denuded at about the same rate as the American.

I think we may assume, without the risk of any great error, that the average rate of sub-aërial denudation during past geological ages did not differ much from the present. The rate at which a country is lowered by sub-aërial denudation is determined (as has been shown, 'Climate and Time,' p. 334) not so much by the character of its rocks as by the sediment-carrying power of its river systems. And this again depends mainly upon the amount of rain-fall, the slope of the ground, and the character of the soil and vegetation covering the surface of the country. And in respect of these we have no reason to believe that the present is materially different from the past. No doubt the average rain-fall during some past epochs might have been greater than at present, but there is just as little reason to doubt that during other epochs it might have been less than now. We may, therefore, conclude that about one foot of rock removed from the general surface of the country in 6000 years may be regarded as not very far from the average rate of denudation during past ages.

But some of the cases we have given of great denudation refer to comparatively small areas, and others to beds which form anticlinal axes, and which, as is well known, denude more rapidly than either synclinal or horizontal beds. We shall therefore, to prevent the possibility of over-estimating the length of time necessary to effect the required amount of denudation, assume the rate to have been double the above, or equal to one foot in 3000 years.

Age of the Stratified Rocks, as determined by the Rate of Denudation.—To lower the country one mile by denudation would therefore require, according to the above rate, about 15 million years; but we have seen that a thickness of rock more than equal to that must have been swept away since the Carboniferous period. For even during the Carboniferous period itself more than a mile in thickness of strata in many places was removed. Again, there can be no doubt whatever that the amount of rock removed during the Old Red Sandstone period was much greater than one mile; for we know perfectly well that over large tracts of country nearly a mile in thickness of rock was carried away between the period of the Lower Old Red Sandstone and the Carboniferous epoch. Further, all geological facts go to show that the time represented by the Lower Old Red Sandstone itself must have been enormous.

Now, three miles of rock removed since the commencement of the Old Red Sandstone period (which in all probability is an under-estimate) would give us 45 million years.

Again, going further back, we find the lapse of time represented by the Silurian period to be even more striking than that of the Old Red Sandstone. The unconformities in the Silurian series indicate that many thousands of feet of these strata were denuded before overlying members of the same great formations were deposited. And again, this immense formation was formed in the ocean by the slow denudation of pre-existing Cambrian continents, just as these had been built up out of the ruins of the still prior Laurentian land. And even here we do not reach the end of the series, for the very Laurentians themselves resulted from the denudation, not of the primary rocks of the globe, but of previously existing sedimentary and probably igneous rocks, of which, perhaps, no recognisable portion now remains.

Few familiar with the facts of geology will consider it too much to assume that the time which had elapsed prior to the Old Red Sandstone was equal to the time which has elapsed since that period. But if we make this assumption, this will give us at least 90 million years as the age of the stratified rocks.

That the foregoing is not an over-estimate of the probable amount of rock removed by sub-aërial denudation during past geological ages will appear further evident from the following considerations: — The mountain ridges of our globe, in most cases, as is well known, have been formed by sub-aërial denudation: they have been carved out of the solid block. They stand two thousand, four thousand, or five thousand feet high, as the case may be, simply because two thousand, four thousand, or five thousand feet of rock have been denuded from the surrounding country. The mountains are high simply because the country has been lowered. But it must be observed that the height which the mountains reach above the surrounding country does not measure the full extent to which the country has been lowered by denudation, because the mountains themselves have also been lowered. The height of the mountains represents merely the extent to which the country has been lowered in excess of the mountains themselves. In the formation of a mountain by denudation, say 3000 feet in height, probably more than 6000 feet of strata may have been removed from the surrouning country. The very fact of a mountain standing above the surrounding country exposes it the more to denudation, and it is certainly not an exaggerated assumption to suppose that whilst the general surface of the country was being lowered 6000 feet by denudation, the mountain itself was at least lowered by 3000 feet.

The very common existence of mountains two or three thousand feet in height formed by sub-aërial denudation, proves that at least one mile must have been worn off the general surface of the country. It does not, of course, follow that the general surface ever stood at an elevation of one mile above the sea-level, since denudation would take place as the land gradually rose. We know that the land was once under the sea, for it was there that it was formed. It is built up out of the materials resulting from the carving out into hill and dale, through countless ages, of a previously existing land, just as this latter had resulted from the destruction of a still older land, and so on in like manner back into the unknown past. We have no means of knowing how often the materials composing the sedimentary rocks may have passed through the process of denudation.*

It has now been proved, by the foregoing very simple and obvious method, that the age of the earth must be far more than 20 or 30 million years. This method, it is true, does not enable us to determine with anything like accuracy the actual age of the globe, but it enables us to determine with absolute certainty that it must be far greater than 20 million years. We have not sufficient data to determine how many years have

* The overlooking of this important point by Mr. Alfred R. Wallace in his determination of the age of stratified rocks ("Island Life," chap. x.) appears to me to vitiate his whole argument. This, I think, has been clearly shown by Mr. T. Mellard Reade, "Geological Magazine," Decade ii. vol. x. pp. 309, 571.

elapsed since life began on the globe, for we do not know the total amount of rock removed by denudation; but we have data perfectly sufficient to show that it began far more than twice 20 million years ago.

But if the present order of things has been existing for more than 20 million years, then the sun must have been illuminating our globe for that period, and, if so, then there must have been some other source than that of gravitation from which the sun derived its energy, for gravitation, as we have seen, could only have supplied the present rate of radiation for about one-half that period.

It is perfectly true, as has been stated, that the length of time that the sun could, by its radiation, have kept the earth in a state fit for animal and vegetable life, must have been limited by the store of energy in the form of heat which it possessed. But it does not follow as a necessary consequence, as is generally supposed, that this store of energy must have been limited to the amount obtained from gravity in the condensation of the sun's mass. The utmost that any physicist is warranted in affirming is simply that it is impossible for him to *conceive* of any other source. His *inability*, however, to conceive of another source cannot be accepted as a proof that there *is* no other source. But the physical argument that the age of our earth must be limited by the amount of heat which could have been received from gravity is in reality based upon this assumption—that, because no other source can be conceived of, there is no other source.

It is perfectly obvious, then, that this mere negative evidence against the possibility of the age of our habitable globe being more than 20 or 30 million years is of no weight whatever when pitted against the positive evidence here advanced, that its age must be far greater.

Now, in proving that the antiquity of our habitable globe must be far greater than 20 or 30 million years, we prove that there must have been some other source in addition to gravity from which the sun derived his store of energy; *and this is the point which I have been endeavouring to reach by this somewhat lengthy discussion.*

CHAPTER XVIII.

THE PROBABLE ORIGIN AND AGE OF THE SUN'S HEAT.—*Continued.*

Not obliged to assume that Gravitation is the only Source of the Sun's Heat.—How the Mass obtained its Temperature.—No Limit to the Amount of Heat which may have been produced.—Age of the Sun in Relation to Evolution.—Note on Sir William Thomson's Arguments for the Age of the Earth.

ARE we really under the necessity of assuming that the sun's heat was wholly, or even mainly, derived from the condensation of his mass by gravity? According to Helmholtz's theory of the origin of the sun's heat by condensation, it is assumed that the matter composing the sun, when it existed in space as a nebulous mass, was not originally possessed of temperature, but that the temperature was given to it as the mass became condensed under the force of gravitation. It is supposed that the heat given out is simply the heat of condensation. But it is quite conceivable that the nebulous mass might have been possessed of an original store of heat previous to condensation.

It is quite possible that the very reason why it existed in such a rarified or gaseous condition was its excessive temperature, and that condensation only began to take place when the mass began to cool down. It seems far more probable that this should have been the case than that the mass existed in so rarefied a condition without temperature. For why should the particles have existed in this separate form when devoid of the repulsive energy of heat, seeing that, in virtue of gravitation, they had such a tendency to approach one another?

It will not do to begin with the assumption of a cold nebulous mass, for, the moment that the mass existed as such, condensation—under the influence of the mutual attraction of its particles—would commence. We must therefore assume either that the mass was created at the moment condensation began, or that, prior to this moment, it existed under some other form. There are few, I think, who would be willing to adopt the former alternative. If we adopt the latter we must then ask the question, In what condition did this mass exist prior to the commencement of condensation? The answer to this question would naturally be that it existed in a condition of excessive temperature, the repulsive force of heat preventing the particles approaching one another. In short, the excessive temperature was the very cause of the nebulous condition.

But if the mass was originally in a heated condition, then in condensing it would have to part not only with the heat of condensation, but also with the heat which it originally possessed.

It is therefore evident that if we admit that the nebulous mass was in a state of incandescence prior to condensation, it will really be difficult to fix any limit either to the age of the sun or to the amount of heat which it may have originally possessed. The 20 million years' heat obtained by condensation may in such a case be but a small fraction of the total quantity possessed by the mass.

How the Mass obtained its Temperature.—The question now arises, By what means could the nebulous mass have become incandescent? From what source could the heat have been obtained? The dynamical theory of heat affords, as was shown several years ago,* an easy answer to this question. The answer is that *the energy in the form of heat possessed by the mass may have been derived from Motion in Space.* Two bodies, each one-half the mass of the sun, moving directly towards each other with a velocity of 476 miles per second, would by their concussion generate in a *single moment* 50 million years' heat. For two bodies of that mass, moving with a velocity of 476 miles per second, would possess 4149×10^{38} foot-pounds of kinetic energy, and this converted into heat by the stoppage of their motion would give out an amount of heat which would cover the present rate of the sun's radiation for a period of 50 million years.†

Why may not the sun have been composed of two such bodies? And why may not the original store of heat possessed by him have all been derived from the concussion of these two bodies? Two such bodies coming into collision with that velocity would be dissipated into vapour and converted into a nebulous mass by such an inconceivable amount of heat as would thus be generated; and when condensation on cooling took place, a spherical mass like that of the sun would result. It is perfectly true that two such bodies could never attain the required amount of velocity by their mutual gravitation towards each other. But there is no necessity whatever for supposing that their velocities were derived from their mutual attraction alone: they might have been approaching each

* "Phil. Mag." for May, 1868.

† It does not necessarily follow, of course, that the two bodies coming into collision should possess equal mass or velocity in order to have their motion of translation converted into heat.

other with the required velocity wholly independent of gravitation.

We know nothing whatever regarding the absolute motion of bodies in Space; and, beyond the limited sphere of our observation, we know nothing even of their relative motions. There may be bodies moving in relation to our system with inconceivable velocity. For anything that we know to the contrary, were one of these bodies to strike our earth the shock might be sufficient to generate an amount of heat that would dissipate the earth into vapour, though the striking body might not be heavier than a cannon-ball. There is, however, nothing very extraordinary in the velocity which we have found would be required to generate the 50 millions years' heat in the case of the two supposed bodies. A comet having an orbit extending to the path of the planet Neptune, approaching so near the sun as to almost graze his surface in passing, would have a velocity of about 390 miles per second, which is within 86 miles of that required.

It must be borne in mind, however, that the 476 miles per second is the velocity at the moment of collision; but more than one-half of this would be derived from the mutual attraction of the two bodies in their approach to each other. Suppose, for simplicity of calculation, each body to be equal in volume to the sun, and of course one-half the density, the amount of velocity which they would acquire by their mutual attraction would be 274 miles per second. Consequently we have to assume an original or projected velocity of only 202 miles per second. And if we assume the original velocity to have been 1700 miles per second, an amount of heat would be generated in a single moment which would suffice for no less than 800,000,000 years. And when we take into consideration the magnitude of the stellar universe, the difference between a motion of 202 miles per second, and one of 1700 miles to a great extent disappears, and the one velocity becomes about as probable as the other. If the original velocity was 676 per second, the total amount of heat generated would suffice for 200 million years at the present rate of radiation.

On former occasions* I expressed it as my opinion that the total quantity of heat possessed by the sun could not probably exceed 100 million years' heat. But if we admit that the heat was derived from Motion in Space, there really does not seem any reason why it may not be double or quadruple that amount.

It will be asked, Where did the two bodies get their motion? It may as well, however, be asked, Where did they get their existence? It is just as easy to conceive that they always existed in motion as to conceive that they always existed at rest. In fact, this is the only way in which energy can remain in a body without dissipation into Space. Under other forms a certain amount of the energy is constantly being transformed into heat which never can be re-transformed back again, but is dissipated into Space as radiant heat. But a body moving in void stellar space will, unless a collision takes place, retain its energy in the form of motion untransformed for ever.

It will perhaps be urged as an objection that we have no experience of bodies moving in space with velocities approaching to anything like 400 or 600 miles per second. A little consideration will, however, show that this is an objection which can hardly be admitted, as we are not in a position to be able to perceive bodies moving with such velocities. No body moving at the rate of 400 miles per second could remain as a member

* "Phil. Mag.," May, 1868. 'Climate and Time,' chap. 21.

of our solar system. Beyond our system, the only bodies visible to us are the nebulæ and fixed stars, and they are visible because they are luminous. But the fixed stars are beyond doubt suns similar to our own; and if we assume that the energy in the form of heat and light possessed by our sun has been derived from Motion in Space, we are hardly warranted in denying that the light and heat possessed by the stars were derived from another source. It is true that the motion of the stars in relation to one another, or in relation to our system (and this is the only motion known to us), is but trifling in comparison to what we even witness in our solar system. But this is what we ought, *à priori*, to expect; for if their light and heat were derived from Motion in Space, like that of our sun, then, like the sun, they must have lost their motion. In fact, *they are suns, and visible because they have lost their motion.* Had not the masses of which these suns were composed lost their motion they would have been non-luminous, and of course totally invisible to us. In short, we only see in stellar space those bodies which, by coming into collision, have lost their motion, for it is the lost motion which renders them luminous and visible.*

* When the foregoing theory of the origin of the sun's heat was advanced, in 1868, I was not aware that a paper on the "Physical Constitution of the Sun and Stars" had been read before the Royal Society by Mr. G. Johnstone Stoney, in which he suggested that the heat possessed by the stars may have been derived from collisions with one another. "If two stars," he says, "should be brought by their proper motion very close, one of three things would happen:—Either they would pass quite clear of one another, in which case they would recede to the same immense distance asunder from which they had come; or they would become so entangled with one another as to emerge from the frightful conflagration which would ensue, as one star; or, thirdly, they would brush against one another, but not to the extent of preventing the stars from getting clear again."

The formation of a sun by collision is an event that would not be likely to escape observation if it occurred within the limits of visibility in space. But such an event must be of very rare occurrence, or the number of stars visible would be far greater than it is. The number of stars registered down to the seventh magnitude, inclusive, is—according to Herschel—somewhere between 12,000 and 15,000, and this is all that can possibly be seen by the naked eye. Now, if we suppose each of them to shine like our sun for (say) 100 million years, then one formed in every 7000 or 8000 years would maintain the present number undiminished. But this is the number included in both hemispheres, so that the occurrence of an event of such unparalleled splendour and magnificence as the formation of a star or rather nebula—for this would be the form first assumed—is what can only be expected to be seen on our hemisphere once in about 15,000 years.

The absence of any historical record of such an event having ever occurred can therefore be no evidence whatever against the theory.

Age of the Sun in relation to Evolution.—One of

In the latter case he considers a double star is formed. Mr. Stoney's paper, though read in 1867, was not published till 1869.

Mr. Herbert Spencer, in his "First Principles" (pp. 532 to 535), has also directed attention to the fact that the stars distributed through space must tend, under the influence of gravity, to concentrate and become locally aggregated. Separate aggregations will be drawn towards one another, and ultimately coalesce. The result will be that the heat evolved by such collisions taking place under the enormous velocities acquired by gravity must have the effect of dissipating the matter of which they are composed into the gaseous state.

Both Mr. Stoney and Mr. Spencer consider the motions of the cosmic masses to be due wholly to gravity, but, as we have seen, gravity alone cannot account for the enormous amount of energy originally possessed by the sun.

the most formidable objections to the theory of evolution is the enormous length of time which it demands. On this point Prof. Haeckel, one of the highest authorities on the subject, in his "History of Creation," has the following:—"Darwin's theory, as well that of Lyell, renders the assumption of immense periods absolutely necessary. . . . If the theory of development be true at all there must certainly have elapsed immense periods, utterly inconceivable to us, during which the gradual historical development of the animal and vegetable proceeded by the slow transformation of species. . . The periods during which species originated by gradual transmutation must not be calculated by single centuries, but by hundreds and by millions of centuries. Every process of development is the more intelligible the longer it is assumed to last."

There are few evolutionists, I presume, who will dispute the accuracy of these statements; but the question arises, does physical science permit the assumption of such enormous periods? We shall now consider the way in which Prof. Haeckel endeavours to answer this question and to meet the objections urged against the enormous lapse of time assumed for evolution.

"I beg leave to remark," he says, "that we have not a single rational ground for conceiving the time requisite to be limited in any way. . . . It is absolutely impossible to see what can in any way limit us in assuming long periods of time. . . . From a strictly philosophical point of view it makes no difference whether we hypothetically assume for this process ten millions or ten thousand millions of years. . . . In the same way as the distances between the different planetary systems are not calculated by miles but by Sirius-distances, each of which comprises millions of miles, so the organic history of the earth must not be calculated by thousands of years, but by palæontological or geological periods, each of which comprises many thousands of years, and perhaps millions or milliards of thousands of years."

Statements more utterly opposed to the present state of modern science on this subject could hardly well be made. Not only have physicists fixed a limit to the extent of time available to the evolutionist, but they have fixed it within very narrow boundaries.

Every one will admit that the organic history of our globe must have been limited by the age of the sun's heat. The extent of time that the evolutionist is allowed to assume depends, therefore, on the answer to the question, What is the age of the sun's heat? And this again depends on the ulterior question, From what source has he derived his energy? The sun is losing heat at the enormous rate of 7,000 horse-power on every square foot of surface. And were it composed of coal its combustion would not maintain the present rate of radiation for 5,000 years. Combustion, therefore, cannot be the origin of the heat.

Gravitation has generally been considered by physicists as the only source from which the sun could have obtained his energy. The contraction theory advocated by Helmholtz is the one generally accepted, but the total amount of work performed by gravitation in the condensation of the sun from a nebulous mass to its present size could only have afforded twenty million years' heat at the present rate of radiation. On the assumption that the sun's density increases towards the centre, a few additional million years' heat might be obtained. But on every conceivable supposition gravitation could not have afforded more than twenty or thirty million years' heat.

Prof. Haeckel may make any assumption he chooses about the age of the sun, but he must not do so in regard to the age of the sun's heat. One who believes it *inconceivable* that matter can either be created or annihilated may be allowed to maintain that the sun existed from all eternity, but few will admit the assumption that our luminary has been losing heat from all eternity.

If 20,000,000 or 30,000,000 years do not suffice for the evolution theory, then either that or the gravitation theory of the origin of the sun's heat will have to be abandoned.

It was proved in the last chapter from geological evidence that the antiquity of our habitable globe must be at least three times greater than it could possibly be had the sun derived its heat simply from the condensation of its mass. This proves that the gravitation theory of the origin of the sun's heat is as irreconcilable with geological facts as it is, according to Haeckel, with those of evolution, and that there must have been some other source, in addition, at least, to gravity, from which the sun derived his store of energy. That other source we have just considered at some length, and found it perfectly adequate.

The theory that the sun's heat was originally derived from motion in space is, therefore, more in harmony with the principles of evolution than the gravitation theory, because it explains how the enormous amount of energy which is being dissipated into stellar space may have existed in the matter composing the sun untransformed during bygone ages, or, in fact, for as far back as the matter itself existed.

Note on Arguments for the Age of the Earth.—Sir William Thomson has endeavoured to prove the recent age of the earth by three well-known arguments of a purely physical nature. The first is based on the age of the sun's heat; the second, on the tidal retardation of the earth's rotation; and the third, on the secular cooling of the earth. Several years ago I ventured to point out some difficulties which appear to me to beset these arguments. They are as follows:—

Argument from the Age of the Sun's Heat.—It will be obvious that, if what has already been advanced in regard to the origin of the sun's heat be correct, it will follow that the argument for the recent age of the earth, based upon the assumption that the sun could have derived its store of heat only from the condensation of its mass, must be wholly abandoned, and that, in so far as this argument is concerned, there is no known limit to the amount of heat which the sun may have possessed, or to the time during which it may have illuminated the earth.

Argument from Tidal Retardation.—It is well known that, owing to tidal retardation, the rate of the earth's rotation is slowly diminishing; and it is therefore evident that, if we go back for many millions of years, we reach a period when the earth must have been rotating much faster than now. Sir William's argument is* that, had the earth solidified several hundred million of years ago, the flattening at the Poles and the bulging at the Equator would have been much greater than we find them to be. Therefore, because the earth is so little flattened, it must have been rotating, when it became solid, at very nearly the same rate as at present. And as the rate of rotation is becoming slower and slower, it cannot be so many millions of years since solidification took place. A few years ago I ventured to point out†

* Trans. Geol. Soc. of Glasgow, vol. iii., p. 1.

† "Nature," August 21, 1872. 'Climate and Time,' p. 335.

what appeared to be a very obvious objection to this argument, viz., that the influence of sub-aërial denudation in altering the form of the earth had been overlooked. It has been proved, as we have seen, that the rocky surface of our globe is being lowered, on an average, by sub-aërial denudation, at the rate of about 1 foot in 6000 years. It follows as a consequence, from the loss of centrifugal force resulting from the retardation of the earth's rotation occasioned by the friction of the tidal wave, that the sea-level must be slowly sinking at the Equator and rising at the Poles. This, of course, tends to protect the polar regions, and expose equatorial regions to sub-aërial denudation. Now, it is perfectly obvious that unless the sea-level at the Equator has, in consequence of tidal retardation, been sinking during past ages at a greater rate than 1 foot in 6000 years, it is physically impossible the form of our globe could have been very much different from what it is at present, whatever may have been its form when it consolidated, because sub-aërial denudation would have lowered the Equator as rapidly as the sea sank. But in equatorial regions the rate of denudation is no doubt much greater than 1 foot in 6000 years, because there the rainfall is greater than in the temperate regions. It has been shown ('Climate and Time,' p. 336) that the rate at which a country is being lowered by sub-aërial denudation is mainly determined not so much by the character of its rocks as by the sediment-carrying power of its river systems. Consequently, other things being equal, the greater the rainfall the greater will be the rate of denudation. We know that the basin of the Ganges, for example, is being lowered by denudation at the rate of about 1 foot in 2300 years; and this is probably not very far from the average rate at which the equatorial regions are being denuded. It is therefore evident that subaërial denudation is lowering the Equator as rapidly as the sea-level is sinking from loss of rotation, and that consequently we cannot infer from the present form of our globe what was its form when it solidified. In as far as tidal retardation can show to the contrary, its form, when solidification took place, may have been as oblate as that of the planet Jupiter. There is another circumstance which must be taken into account. The lowering of the Equator, by the transference of materials from the Equator to the higher latitudes, must tend to increase the rate of rotation, or, more properly, it must tend to lessen the rate of tidal retardation.

The argument may be shown to be inconclusive from another consideration. The question as to whether the earth's axis of rotation could ever have changed to such an extent as to have affected the climate of the Poles, a few years ago excited a good deal of attention. The subject has been investigated with great care by Professor Houghton,* Mr. George Darwin,† the Rev. J. F. Twisden,‡ and others, and the general result arrived at may be expressed in the words of Mr. G. Darwin:—"If the earth be quite rigid no re-distribution of matter in new continents could ever have caused the deviation of the Pole from its present position to exceed the limit of about 3°."

Mr. Darwin has shown that, in order to produce a displacement of the pole to the extent of only 1° 46, an area equal to one-twentieth of the entire surface of the globe would have to be elevated to the height of two miles. The entire continent of Europe elevated

* Proc. Roy. Soc., vol. xxvi., p. 51.

† Proc. Roy. Soc., vol. xxv., p. 328.

‡ Paper read before the Geological Society, February 21st, 1877.

two miles would not deflect the Pole much over half a degree. Assuming the mean elevation of the continents of Europe and Asia to be 1000 feet, Professor Houghton calculates that their removal would displace the Pole only 199·4 miles.

It may now be admitted as settled that if the earth be perfectly rigid, the climate of our globe could never possibly have been affected by any change in the axis of rotation. But it is maintained that, if the earth can yield as a whole so as to adapt its form to a new axis of rotation, the effects may be cumulative, and that a displacement of the Pole as much as 10° or 15° is possible.

But then if the earth be able to adapt its form to a *change in the axis of rotation,* there is no reason why it may not be able to adapt its form to a *change in the rate of rotation,* and, if so, the flattening at the Poles and the bulging at the Equator would diminish as the rate of rotation diminished, even supposing there were no denudation going on.

Argument from the Secular Cooling of the Earth.— The earth, like the sun, is a body in the process of cooling, and it is evident that if we go back sufficiently far we shall reach a period when it was in a molten condition. Calculating by means of Fourier's mathematical theory of the conductivity of heat, Sir William Thomson has endeavoured to determine how many years must have elapsed since solidification of the earth's crust may have taken place. This argument is undoubtedly the most reliable of the three. Nevertheless, the data on the subject are yet very imperfect, so that no very definite result can be arrived at by this means as to the actual age of the earth. In fact, this is obvious from the very wide limits assigned by him within which solidification probably took place.

"We must," quoting Sir William's own words on the subject, "allow very wide limits on such an estimate as I have attempted to make; but I think we may, with much probability, say that the consolidation cannot have taken place less than 20,000,000 years ago, or we should have more underground heat than we actually have,—nor more than 400,000,000 years ago, or we should not have so much as the least observed underground increment of temperature. That is to say, I conclude that Leibnitz's epoch of 'emergence' of the 'consistentur status' was probably within these dates."*

* Trans. Roy. Soc. of Edinburgh, vol. xxiii., p. 161.

CHAPTER XIX.

THE PROBABLE ORIGIN OF NEBULÆ.

Motion in space as a source of heat.—Reason why nebulæ occupy so much space.—Reason why nebulæ are of such various shapes.—Reason why nebulæ emit such feeble light.—Heat and light of nebulæ cannot result from condensation.—The gaseous state the first condition of a nebula. — Star-clusters. — Objections considered.

THE object of the present chapter is to examine the bearings of the modern science of energy on the question of the origin of nebulæ, and in particular to consider the physical cause of the dispersion of matter into stellar space in the nebulous form. In doing so, I have studiously avoided the introduction of mere hypotheses and principles not generally admitted by physicists. These remarks may be necessary, as the heading to the present chapter might otherwise lead to the belief that it is on a speculative subject lying outside the province of the physicist.

The question of the origin of nebulæ is simplified by the theory, now generally received, that stars are suns like our own, and that nebulæ are in all probability stars in process of formation. The problem will therefore be most readily attacked by considering, first, the origin of our sun, as this orb, being the one most accessible to us, is that with which we are best acquainted.

By the origin of the sun I do not, of course, mean the origin of the matter constituting the sun—this being an inquiry with which the physicist has nothing whatever to do—but simply its origin as a sun, *i.e.*, as a source of light and heat. Our first question must therefore be, What is the origin of the sun's heat? From what source did he derive that enormous amount of energy which, in the form of heat, he has been dissipating into space during past ages? Difficult as the question at first sight appears to be, it is yet simplified and brought within very narrow limits, as was shown in the last chapter, when we remember that there are only two conceivable sources. The sun must have derived his energy either from *Gravitation*, or from that other source which has just been considered, *Motion in Space*.* All other sources of energy put together could not have supplied our luminary with one thousandth part of that which he has possessed. We are therefore compelled to attribute the sun's heat to one or other of these two, or to give up the whole inquiry as utterly hopeless. The important difference between the two, as was shown, is that the store of energy derivable from Gravitation could not possibly have exceeded 20 to 30 million years' supply of heat at the present rate of radiation; whereas the store derivable from Motion in Space, depending on the *rate* of that motion, may conceivably have amounted to any assignable quantity. Thus a mass equal to that of the sun, moving with a velocity of 476 miles per second, possesses in virtue of that motion energy sufficient, if converted into heat, to cover the present rate of the

* A new theory of the sun was advanced a few years ago by Dr. C. W. Siemens, Proc. Roy. Soc., vol. 33, p. 389. According to this theory the greater part of the energy in the form of heat radiated from the sun is arrested and returned over and over again to that luminary. By this means it is concluded the radiation of the sun can be maintained to the remotest future. The theory, I fear, is not likely to gain acceptance among physicists.

sun's radiation for 50 million years. Twice that velocity would give 200 million years' heat; four times that velocity would give 800 million years' heat, and so on without limit.

It is, however, not enough that we should have in the form of motion in space energy sufficient. We must have a means of converting this motion into heat—of converting motion of translation into molecular motion. To understand how this can be effected, we simply require the conception of *Collision.* Two bodies moving towards each other will have their motion of translation converted into molecular motion (heat) by their encounter.

To which of these two causes must we attribute the Sun's heat? It is certain that gravitation must have been a cause; and if we adopt the nebular hypothesis of the origin of our solar system, then from 20 to 30 million years' heat may thus be accounted for. But we know from geological evidence that the sun has been dissipating his light and heat at about the present rate for a much longer period. In Chapter XVI. the geological evidence for the age of the earth was discussed at considerable length, and it was pointed out that the time which has elapsed since life began on the globe cannot have been less than 60 million years. Although we have good grounds for believing that 60 million years have elapsed since life began on the globe, yet the lapse of time may really have been very much longer. We are justified, therefore, in concluding that our globe has been receiving from the sun for the past 60 million years an amount of light and heat daily not very sensibly less than at present. This shows that gravitation alone will not explain the origin of the sun's heat, and that a far more effective cause must be found. Now the only other conceivable cause exceeding that of gravity is, of course, motion in space.

If the gravitation theory fails to explain the origin of the sun, it fails yet more decidedly to account for the nebulæ. In fact it does not attempt any explanation of the origin of the latter; for it begins by *assuming* their existence, and not only so, but that they are in *process of condensation*. This must be the case, because the theory in question assumes that the particles of a nebulous mass have, in virtue of gravity, a mutual tendency to approach one another; and it cannot tell us how this tendency could exist without producing its effect. The advocates of the theory are not at liberty to call in the aid of heat in order to explain why the particles are not mutally approaching; because it is this mutual approach which, according to the theory, produces the heat, and of course without such approach no heat could be generated. A nebulous mass with a tendency to condensation could not have existed from eternity as such; but what the previous condition of a nebula was, and how it came to assume its present state, the gravitation theory cannot say. It begins with a star or sun in process of formation, but does not help us to understand how the process of formation commenced.

It is quite otherwise, however, with the other theory. This latter does not, like the former, begin by assuming the existence of a nebulous mass; on the contrary, it goes back to the very commencement of physical inquiry, to the very point where physical investigation takes its rise, and beyond which we cannot penetrate. The only assumption it makes is that of the existence of matter and motion—if indeed this can be called an assumption. How matter and motion began to be, whether they were eternal or were created, are questions wholly beyond the domain of the physicist. The theory takes for a fact the existence of stellar masses in a state of motion; and its advocate is not required, as a physicist, to account for the existence either of those masses or of their motions. Neither is it necessary for him to advance any hypothesis to show how the masses came into collision; for unless we are to assume that all stellar masses are moving in one direction and with uniform velocity (a supposition contrary to known facts), then collisions must occasionally take place. The chances are that stellar masses are of all sizes, moving in all directions and with all velocities. We have here therefore, without any hypothesis, all the conditions necessary for the origin of nebulæ. Take the case of the origin of the nebulous mass out of which our sun is believed to have been formed. Suppose two bodies, each one half the mass of the sun, approaching each other directly at the rate of 476 miles per second (and there is nothing at all improbable in such a supposition), their collision would transform the whole of the motion into heat affording an amount sufficient to supply the present rate of radiation for 50 million years. Each pound of the mass would, by the stoppage of the motion, possess not less than 100,000,000,000 foot-pounds of energy transformed into heat, or as much heat as would suffice to melt 90 tons of iron or raise 264,000 tons 1° C. The whole mass would be converted into an incandescent gas, with a temperature of which we can form no adequate conception. If we assume the specific heat of the gaseous mass to be equal to that of air (viz. ·2374), the mass would have a temperature of about 300,000,000° C., or more than 140,000 times that of the voltaic arc.

Reason why Nebulæ occupy so much space.—It may be objected that enormous as would be such a temperature, it would nevertheless be insufficient to expand the mass against gravity so as to occupy the entire space included within the orbit of Neptune. To this objection it might be replied, that if the temperature in question were not sufficient to produce the required expansion, it might readily have been so if the two bodies before encounter be assumed to possess a higher velocity, which of course might have been the case. But without making any such assumption, the necessary expansion of the mass can be accounted for on very simple principles. It follows in fact from the theory, that the expansion of the gaseous mass must have been far greater than could have resulted simply from the temperature produced. by the concussion. This will be obvious by considering what must take place immediately after the encounter of the two bodies, and before the mass has had sufficient time to pass completely into the gaseous condition. The two bodies coming into collision with such enormous velocities would not rebound like two elastic balls, neither would they instantly be converted into vapour by the encounter. The first effect of the blow would be to shiver them into fragments, small indeed as compared with the size of the bodies themselves, but still into what might be called in ordinary language immense blocks. Before the motion of the two bodies could be stopped, they would undoubtedly interpenetrate each other; and this, of course, would break them up into fragments. But this would only be the work of a few minutes. Here, then, we should have all the energy of the lost motion existing in these blocks as heat (molecular motion), while they were still in the solid state; for as yet they would not have had sufficient time to assume the gaseous condition. It is obvious, however, that the greater part of the heat would exist on the surface of the blocks (the place receiving the greatest concussion), and would continue there while the blocks retained their solid condition. It is difficult in imagination to realize what the temperature of the surfaces would be at this moment. For, supposing the heat were uniformly distributed through the entire mass, each pound, as we have already seen, would possess 100,000,000,000 foot-pounds of heat. But as the greater part of the heat would at this instant be concentrated on the outer layers of the blocks, these layers would be at once transformed into the gaseous condition, thus enveloping the blocks and filling the interspaces. The temperature of the incandescent gas, owing to this enormous concentration of heat, would be excessive, and its expansive force inconceivably great. As a consequence the blocks would be separated from each other, and driven in all directions with a velocity far more than sufficient to carry them to an infinite distance against the force of gravity were no opposing obstacle in their way. The blocks, by their mutual impact, would be shivered into smaller fragments, each of which would consequently become enveloped in incandescent gas. These smaller fragments would in a similar manner break up into still smaller pieces, and so on until the whole came to assume the gaseous state. The general effect of the explosion, however, would be to disperse the blocks in all directions, radiating from the centre of the mass. Those towards the outer circumference of the mass, meeting with little or no obstruction to their onward progress, would pass outwards into space to indefinite distances, leaving in this manner a free path for the layers of blocks behind them to follow in their track. Thus eventually a space, perhaps twice or even thrice that included within the orbit of Neptune, might be filled with fragments by the time the whole had assumed the gaseous condition.

It would be the suddenness and almost instantaneity with which the mass would receive the entire store of energy, before it had time even to assume the molten, far less the gaseous condition, which would lead to such fearful explosions and dispersion of the materials. If the heat had been gradually applied, no explosions, and consequently no dispersion, of the materials would have taken place. There would first have been a gradual melting; and then the mass would pass by slow degrees into vapour, after which the vapour would rise in temperature as the heat continued, until it became possessed of the entire amount. But the space thus occupied by the gaseous mass would necessarily be very much smaller than in the case we have been considering, where the shattered materials were first dispersed into space before the gaseous condition was assumed.

Reason why Nebulæ are of such various Shapes.—The latter theory accounts also for the various and irregular shapes assumed by the nebulæ; for although the dispersion of the materials would be in all directions, it would, according to the law of chances, very rarely take place uniformly in all directions. There would generally be a greater amount of dispersion in certain directions, and the materials would thus be carried along various lines and to diverse distances; and although gravity would tend to bring the widely scattered materials ultimately together into one or more spherical masses, yet, owing to the exceedingly rarified condition of the gaseous mass, the nebulæ would change form but slowly.

Reason why Nebulæ emit such feeble Light.—The feeble light emitted by nebulæ follows as a necessary result from the theory. The light of nebulæ is mainly derived from glowing hydrogen and nitrogen in a gaseous condition; and it is well known that these gases are exceedingly bad radiators. The oxyhydrogen flame, though its temperature is only surpassed by that of the voltaic arc, gives nevertheless a light so feeble as scarcely to be visible in daylight. Now, even supposing the enormous space occupied by a nebula were due to excessive temperature, the light emitted would yet not be intense were it derived from nitrogen or hydrogen gas. The small luminosity of nebulæ, however, is due to a different cause. The enormous space occupied by those nebulæ is not so much owing to the heat which they possess, as to the fact that their materials were dispersed into space before they had time to pass into the gaseous condition; so that, by the time this latter state was assumed, the space occupied was far greater than was demanded either by the temperature or the amount of heat received.

If we adopt the nebular hypothesis of the origin of our solar system, we must assume that our sun's mass, when in the condition of nebula, extended beyond the orbit of the planet Neptune, and consequently filled the entire space included within that orbit. Supposing Neptune's orbit to have been its outer limit, which it evidently was not, it would nevertheless have then occupied 274,000,000,000 times the space that it does at present. We shall assume, as before, that 50 million years' heat was generated by the concussion. Of course there might have been twice, or even ten times that quantity; but it is of no importance what number of years is in the meantime adopted. Enormous as 50 million of years' heat is, it yet gives, as we shall presently see, only 32 foot-pounds for each cubic foot. The amount of heat due to concussion being equal, as before stated, to 100,000,000,000 foot-pounds for each pound of the mass, and a cubic foot of the sun at his present density of 1·43 weighing 89 lbs., each cubic foot must have possessed 8,900,000,000,000 foot-pounds. But when the mass was expanded to occupy 274,000,000,000 times more space, which it would do when it extended to the orbit of Neptune, the heat possessed by each cubic foot would then amount to only 32 foot-pounds.

In point of fact, however, it would not even amount to that; for a quantity equal to upwards of 20 million years' heat would necessarily be consumed in work against gravity in the expansion of the mass; all of which would, of course, be given back in the form of heat as the mass contracted. During the nebulous condition it would not exist as heat, so that only 19 foot-pounds out of the 32 foot-pounds generated by concussion would then exist as heat. The density of the nebula would be only $\frac{1}{16248160}$ that of hydrogen at ordinary temperature and pressure. The 19 foot-pounds of heat in each cubic foot would nevertheless be sufficient to maintain an excessive temperature; for there would be in each cubic foot only $\frac{1}{440000}$ of a grain of matter. But although the *temperature* would be excessive, the *quantity* both of light and heat in each cubic foot would of necessity be small. The heat being only $\frac{1}{71}$ of a thermal unit, the light emitted would certainly be exceedingly feeble, resembling very much the electric light in a vacuum-tube.

Heat and Light of Nebulæ cannot result from Condensation.—The fact that nebulæ are not only self-luminous but indicate the existence of hydrogen and nitrogen in an incandescent condition proves that they must possess a considerable temperature. And it is scarcely conceivable that the temperature could have been derived from the condensation of their masses.

When our sun was in the nebulous condition it no doubt was self-luminous like other nebulæ, and doubtless would have appeared, if seen from one of the fixed stars, pretty much like other nebulæ as viewed from our earth. The spectrum would no doubt have revealed in it the presence of incandescent gas. At all events we have no reason to conclude that our nebula was in this respect an exception to the general rule, and essentially different from others of the same class. The heat which our nebula could have derived from condensation up to the time that Neptune was formed, no matter how far the outer circumference of the mass may originally have extended beyond the orbit of that planet, could not have amounted to over $\frac{1}{7000000}$ of a thermal unit for each cubic foot; and the quantity of light given out could not possibly have rendered the mass visible. Consequently the heat and light possessed by the mass must have been derived from some other source than that of gravity.

We have further evidence that the heat and light of nebulæ cannot have been derived from condensation. If there be any truth, as there doubtless is, in Mr. Lockyer's view of the evolution of the planets, then the nebulæ out of which these bodies were evolved must have originally possessed a very high temperature—a temperature so high, indeed, as to produce perfect chemical dissociation of the elements. In short, "the temperature of the nebulæ," as Mr. Lockyer remarks,* "was then as great as the temperature of the sun is now." Mr. Lockyer's theory is that the metals and the metalloids, owing to excessive temperature, existed in the nebulous mass uncombined—the metals, owing to their greater density, assuming the central position, and the metalloids keeping to the outside. The denser

* "Why the Earth's Chemistry is as it is," p. 55, 1877.

the metal the nearer would its position be to the centre of the mass, and the lighter the metalloid the nearer to the outside. As a general rule the dissociated elements would arrange themselves according to their densities; and it is for this reason, he considers, that the outer planets Neptune, Uranus, Saturn, and Jupiter, are less dense than the inner planets, since they must have been formed chiefly of metalloids, while the inner and more dense planets would consist chiefly of metallic elements.

"The hypothesis," says Mr. Lockyer, "is almost worthless unless we assume very high temperatures, because unless you have heat enough to give perfect dissociation, you will not have that sorting-out which always seems to follow the same law." But the heat which produced this dissociation previous to the formation of the planets could not have been derived from the condensation of the nebula; for the quantity so derived prior to the existence of the outermost planet must have been infinitesimal indeed. The heat existing in the nebula previous to condensation must have come from some other source; and we can conceive of no other save that which we have been considering.

The Gaseous State the first Condition of a Nebula.— If the foregoing be the true explanation of the origin of nebulæ, it will follow that the gaseous state will in most cases be the first or original condition, and that a nebula giving a continuous spectrum will only be found after it has condensed to a considerable extent.

The irresolvable nebulæ which exhibit bright lines, in all probability consist, as Mr. Huggins maintains, of glowing gas without anything solid in them. In short, they are nebulæ in their first stage of development, and have not as yet condensed sufficiently to become possessed of nuclei. If we adopt the generally accepted nebular hypothesis, I cannot understand how we can consistently deny the existence of gaseous nebulæ; for, according to the nebular hypothesis, the central nucleus which constitutes a sun or star, and which exhibits a continuous spectrum, was formed by condensation as surely as the planets or the satellites have been. Were we to go back sufficiently far in the past, we should come to a time when not only our globe but the sun himself consisted of gaseous matter only. If we admit this, then why not also admit that there may be nebulæ at the present time in a condition similar to what our sun must formerly have been.

The gaseous condition of the nebulæ seems to follow as a consequence from Mr. Lockyer's theory. For, in order that the materials in the formation of a sun or star may arrange themselves according to their densities, dissociation is requisite; but there can be no dissociation except in the gaseous condition.

Star - Clusters.—The wide - spread and irregular manner in which the materials would in many cases be distributed through space after collision, would prevent a nebula from condensing into a single mass. Subordinate centres of attraction, as was long ago shown by Sir William Herschel (in his famous memoir on the formation of stars*), would be established, around which the gaseous particles would arrange themselves and gradually condense into separate stars, which would finally assume the condition of a cluster.

Binary, Triple, and Multiple systems of stars will of course be accounted for in a similar manner.

It is conceivable that it may sometimes happen that by the time the materials are broken up and dissipated into space, there may not be sufficient heat left to convert the fragments into vapour. In this case we

* "Phil. Trans." for 1811.

should have what Professor Tait has suggested, a nebula consisting of "clouds of stones." But such nebulæ must be of rare occurrence.

Objections considered.—On a former occasion I considered one or two anticipated objections to the theory that stellar light and heat were derived from motion in space. But as these objections have since been repeatedly urged by physicists both in this country and in America, I shall again briefly refer to them.

Objection 1st. "The existence of such non-luminous bodies as the theory assumes is purely conjectural, as no such bodies have ever been observed." In reply, it is just as legitimate an inference that there are bodies in stellar space not luminous as that there are luminous bodies in space not visible. We have just as good evidence for believing in the existence of the one as we have in the existence of the other. Bodies in stellar space can only be known through the eye to exist. If they are not luminous, they of course cannot be seen. But we are not warranted on that account to suppose that they do not exist, any more than we have to suppose that stars do not exist which are beyond the reach of our vision. We have, however, positive evidence that there are bodies in space non-luminous, as the meteorites and planets for example. The stars are beyond doubt suns like our own; and we cannot avoid the inference that, like our sun, they are surrounded by planets. If so, then we have to admit that there are far more bodies in stellar space non-luminous than luminous. But this is not all: the stars no more than our sun can have been dissipating their light and heat during all past ages; their light and heat must have had a beginning; and before that they could not be luminous. Neither can they continue to give out light and heat eternally; consequently, when their store of energy is exhausted, they will be non-luminous again. Light and heat are not the permanent possession of a body. A body may retain its energy in the form of motion undiminished and untransformed through all eternity, but not so in the form of heat and light. These are forms of energy which are being constantly dissipated into space and lost in so far as the body is concerned.

The conclusion to which we are therefore led is that there are in all probability bodies in stellar space which have not yet received their store of light and heat, while there are others which have entirely lost it. The stars are probably only those stellar masses which, having recently had an encounter, have become possessed of light and heat. They have gained in light and heat what they have lost in motion, but they have gained a possession which they cannot retain, and when it is lost they become again what they originally were—dark bodies.

2nd. "We have no instances of stellar motions comparable with those demanded by the theory." A little consideration will show that this is an objection which, like the former, can hardly be admitted. No body, of course, moving at the rate of 400 miles per second, could remain a member of our solar system; and beyond our system the only bodies visible are the nebulæ and fixed stars; and they are according to the theory visible, because, like the sun, they have lost their motion—the lost motion being the origin of their light and heat. Their comparatively small velocities are in reality evidence in favour of the theory than otherwise; for had the stars been moving with excessive velocities, this would have been adduced as proof that their light and heat could not have been derived from motion lost, as the theory assumes.

3rd. "If suns or stars have been formed by collision of bodies moving in space, proper motion can be none other than the unused and unconverted energy of the original components. And as stellar bodies are likely of all sizes, and moving with all manner of velocities, it must often happen, from the unequal force of the impinging masses, that a large proportion of the original motion must remain unconverted into heat. Consequently, some of the stars ought, according to the theory, to possess great velocities—which is not the case, as none of the stars have a motion of more than 30 or 40 miles per second."

I freely admit that, if it could be proved that none of the stars have a proper motion of more than 30 or 40 miles per second, it would at least be a formidable difficulty in the way of accepting the theory. For it would indeed be strange that, amidst all the diversity of dimensions of heavenly bodies, it should invariably happen that the resultant movement of the combined masses should be reduced to such comparatively insignificant figures. But something more definite must yet be known in reference to the motion of the stars before this objection can be urged.

All that we are at present warranted to assume is simply that, of the comparatively few stars whose rate of motion has been properly measured, none have a greater velocity than 30 or 40 miles per second, while nothing whatever is known with certainty as to the rate of motion of the greater number of the stars.

There seems to be a somewhat prevailing misapprehension regarding the extent of our knowledge of stellar motions. Before we can ascertain the rate of motion of a star from its angular displacement of position in a given time, we must know its absolute distance. But it is only of the few stars which show a well-marked parallax that we can estimate the distance; for it is now generally admitted that there is no relation between the apparent magnitude and the real distance of a star. All that we know in regard to the distances of the greater mass of the stars is little else than mere conjecture. Even supposing we knew the absolute distance of a star and could measure its amount of displacement in a given time, still we could not be certain of its rate of motion unless we knew that it was moving directly at right angles to the line of vision, and not at the same time receding or advancing towards us; and this we could not determine by mere observation. The rate of motion, as determined from its observed change of position, may be, say, only twenty miles a second, while its actual velocity may be ten times that amount.

By spectrum-analysis it is true we can determine the rate at which a star may be advancing or receding along the line of sight independently of any knowledge of its distance. But this again does not give us the actual rate of motion, unless we are certain that it is moving directly to or from us. If it is at the same time moving transversely to the observer, its actual motion may be more than a hundred miles per second, while the rate at which it is receding or advancing, as determined by spectrum-analysis, may not be 20 miles a second. But in many cases it would be difficult to ascertain whether the star had a transverse motion or not. A star, for example, 1000 times more remote than *a* Centauri (that is, twenty thousand billion miles), though moving transversely to the observer at the enormous rate of 100 miles per second, would take upwards of 30 years to change its position so much as 1″, and 1800 years to change its position 1′; in fact we should have to watch the star for a generation or two before we could be certain whether it was changing its position or not. And even after we had found with certainty that the star was shifting, and this at the rate of 1′ in 1800 years, we could not, without a knowledge of its distance, express the angle of displacement in miles. But from the apparent magnitude or brilliancy of the star, we could not determine whether its distance was 10 times, 100 times, or 1000 times that of *a* Centauri; and consequently we could form no conjecture as to the actual velocity of the star. If we assumed its distance to be 10 times that of *a* Centauri, this would give a transverse velocity of one mile per second. If we assumed its distance to be 100 times that of *a* Centauri, this would give 10 miles a second as the velocity, and if 1000 times, the velocity of course would be 100 miles per second.

As there are but few of the stars which show a measurable parallax, and we have no other reliable method of estimating their distances, * it follows that in reference to the greater number of the stars, neither by spectrum-analysis nor by observation of their change of position can we determine their velocities. There does not, therefore, appear to be the shadow of a reason for believing that none of the stars has a motion of over 30 or 40 miles per second: for any thing that at present is known to the contrary, many of them may possess a proper motion enormously greater than that.

There is, however, an important point which seems to be overlooked in this objection, viz., that, unless the greater part of the motion of translation be transformed into heat, the chances are that no sun star will be formed. It is necessary to the formation of a sun

* It is true that we may one day be able to determine by spectrum-analysis the distance of some of the binary stars; but as yet this method has not been applied with success.

which is to endure for millions of years, and to form the centre of a planetary system like our own, that the masses coming into collision should be converted into an incandescent nebulous mass. But the greater the amount of motion left unconverted into heat, the less is the chance of this condition being attained. A concussion which would leave the greater part of the motion of translation untransformed, would be likely as a general rule to produce merely a temporary star, which would blaze forth for a few years, or a few hundred years, or perhaps a few thousand years and then die out. In fact we have had several good examples of such since the time of Hipparchus. Now, although it may be true that, according to the law of chances, collisions producing temporary stars must be far more numerous than those resulting in the formation of permanent stars, nevertheless the number of those temporary stars observable in the heavens may be perfectly insignificant in comparison with the number of permanent stars. Suppose there were as many as one hundred temporary stars formed for one permanent, and that on an average each should continue visible for 1000 years, there would not at the present moment be over half-a-dozen of such stars visible in the heavens.

4th. "Such collisions as the theory assumes are wholly hypothetical; it is extremely improbable that two cosmical bodies should move in the same straight line; and of two moving in different lines, it is improbable that either should impinge against the other." I reply, if there are stellar masses moving in all directions, collisions are unavoidable. It is true they will be of rare occurrence; but it is well that it is so; for if they had been frequent the universe would be in a blaze, and its store of energy soon converted into heat.

INDEX.

ACER *pseudo-platinus*, 189
" Adamshina," 181, 193
Aërial currents, heat of, 23
,, currents, misapprehensions regarding, 23
Age of sun, 282
,, of sun in relation to evolution, 288
Air at the Equator, why not hotter in January, 59-63
Alnaster fruticosus, 183
,, in Siberia, 184
Alps, elevation of, 174
Alternations of climate, evidence of, 160
Anodonta anatina, 184
Antarctic ground under ice, temperature of, 226
,, land flatness proved by stratification of ice, 229
,, continent, ice of, 64
Antarctic ice, Captain Sir F. J. O. Evans' Theory of, 77
,, formed on flat land, 72-78
,, and geographical conditions, 110
,, on the melting of, 111
,, present extension, cause of, 113
,, temperature of, 213
,, compression and friction in relation to, 218
,, pressure cannot raise its temperature, 231
,, influence of eccentricity on, 111
Antarctic ice, thickness of, proved by icebergs, 211
,, absolutely free from stones or gravel, 230
,, thickness of due to difficulty of removal, 239
,, rate of motion of, 237
,, cause of motion of, 240
,, estimate of area covered by, 238
,, conclusions established regarding, 241
Antarctic icebergs, peculiar formation, 70-73
,, an important consequence of, 74
,, Sir Wyville Thomson on, 203
Antarctic ice-cap, section across, 243
,, shallowness of, illustrated, 243
,, motion of, due to piling up of ice at centre, 234
Antarctic ice-sheet, physical conditions of, 202
,, thickness of, not limited by pressure, 224
,, annual discharge of icebergs from, 238
,, temperature determined mainly by that of its surface, 220

Antarctic ice-sheet, little affected by transmitted heat, 218
,, mean temperature of, 221
,, interior of, warmer than surface, 217
Antarctic regions, 69
,, Sir Wyville Thomson on, 203
,, Wilkes and Ross on snowfall of, 79
Aphelion, influence of winter in, 39
,, effect of winter solstice in, 83
Aqueous vapour as a "trap" for heat, 31
,, as a "screen," 32
,, an important result from, 34
,, influence on snow-line, 47
Arctic America, ice-cliffs of, 192
,, during interglacial times, 192
,, why interglacial deposits so seldom found in, 193
Arctic flora, Prof. J. Geikie on distribution of, 97
Arctic mild climates, Sir W. Thomson on, 148-150
,, Mr. A. R. Wallace on, 153-157
,, accounted for, 147
Atlantic, North: its three sources of heat, 259
Atmosphere, heat cut off by, 37
Awamka, shells at the mouth of, 183

BALTIC filled with ice during Glacial Epoch, 247
Banks's Land, wood found at, 193
Belgium, mammoth remains found at, 186
Belt, Mr., cited, 184
Berendt, Prof., on German interglacial beds, 133
Bodies, non-luminous in space, 310
Boulder-clays of Germany, 133
Brandt, M., cited, 182
Boussingault on lowering of melting point of ice, 215
Brown, Dr. Robert, on Greenland ice, 64-66
,, cited, 67, 69
Buried forests, climatic condition of, 117-119
,, under Carse-clays, 117

CAITHNESS, glaciation of, by land-ice, 170
Canary Islands, 188
,, laurel, 188
Canstadt, in Würtemberg, tufa of, 189, 190
Carse-clays of Scotland, 115
,, "flour of rock," 115
,, winter temperature of, period, 115
,, sea-level higher than at present, 117
Cataclysmic Theories of geological climate, 11
Cause, Sir W. Hamilton's definition of, 103
a Centauri, illustrations from distance of, 314
Chambers, Robert, on moraines on raised beach, 116
Chart of path of ice, 133
Chukchi, peninsula of, mammoth found there, 179
Climate, evidence of alterations of, 160
Cold, storage of, 83
Collisions between stellar bodies, effects of described, 302
Combustion theory of sun's heat untenable, 264
Conduction, influence of, on ice, 216
Continental ice must of necessity be thickest at centre, 234
Continental ice, difference of from a glacier, 202
Continental ice at present on low lands, 88
Credner, Prof., on German interglacial beds, 133
Cyclas calyculata, 183
Cyrena fluminalis, 131, 134, 184

DANA, Prof., on areas most glaciated, 56
,, on thickness of ice in America, 212
Darwin, Prof. George, on obliquity of ecliptic, 4
,, on change of earth's axis, 294
Denmark, mammoth remains found in, 186
Denudation as a measure of geological time, 274
,, in relation to age of earth, 267
,, rates of, 274, 276
,, mountains carved out by, 278
,, enormous, details of, 271, 272
,, effect of tidal retardation neutralised by, 293
,, before and after dislocations, 273
Diagram of section across Antarctic ice-cap, 243
Distribution in Arctic regions, 197
,, flora and fauna, theory of, 197
,, influence of Gulf-stream on, 198
Dove on mean temperature of two hemispheres, 60
Dudino, on limit of wood, 182
Dulong and Petit's Law, 22
,, formula, 260, 145

EARN, carse-clays of, 195
Earth, note on arguments for age of, by Sir W. Thomson, 292
,, axis of rotation of, unchanged, 294
,, axis of rotation, changes of, 5
Eccentricity, effect of, 15
,, formula of, 38
,, Tables of, 38, 39
,, very high not required to produce glaciation, 53
Eccentricity, alone cannot produce glaciation, 165
,, a misapprehension regarding, 92
,, influence during tertiary period, 157
,, influence on tertiary climate, 160
,, gives dates for glacial periods, 174
Ecliptic, obliquity of, 115
Elephas antiquus, 134
Enderby Land under glaciation, 113
Energy, absolute store of, determines age of sun, 280
England, inter-glacial beds of, 131
,, south of, difficulty of finding proofs of glaciation in, 168
Eocene glacial deposits of Switzerland, 173
,, period, date of, 174
Equator, heat received at, to that at the poles, 143
,, and poles, what ought to be the difference, 146
Equatorial water, influence on polar climate, 165
Erratic blocks, on absence of in Tertiary strata, 168
Europe, condition during glacial epoch, 129, 132
,, north-western, ice-sheet of, 246
,, N.W., during the glacial epoch, 74
Evaporation and eccentricity, 23
Evans, Capt. Sir F. J. O., on Antarctic ice, 78
Evolution, age of sun in relation to, 288
,, of the planets, Mr. Lockyer on, 307
Evolutionists' demands for time by, 290

FARADAY, Prof., theory of regelation, 250
Faröe Islands, land-barrier, 98
,, not overridden by ice-sheet, 133

Faröe Islands, distribution of flora in, 197
Faults, evidence from, of the earth's age, 268
,, large, details of, 268-270
Ferrell, Mr. W., on temperature of the two hemispheres, 36
Fitzroy, Admiral, on measurements of icebergs, 210
"Flysch" of Switzerland, 173
Fogs, influence on climate, 42, 44, 120
Formations, time of, as measured by denudation, 277
Forest-beds, probable date of, 117
,, under Carse-clays, 117
,, sea lower than at present, 117
Forth, Carse-clays of, 115
Frankland, Prof., on cause of glacial epoch, 2
France, mammoth remains found in, 186
Fuego, why the snow is permanent, 51

GARDNER, Mr. J. Starkie, on Tertiary fossil floras, 161
,, refutation of Mr. Wood's theory, 163
,, cited, 161, 163, 174
Gases, radiation of heat through, 262
Gastaldi, M., on glacial deposits in Miocene beds, 172
Geikie, Prof. A., on large faults, 269
Geikie, Prof. J., on climate of Pleistocene times, 127
,, on post-glacial elevation, 97
,, on distribution of Arctic flora, 197
,, on land-connection between Europe and Faröes, 197
,, on tufa of Europe, 187
,, on Carse-clays, 114, 115
Geike, Prof. J., on slopes down which ice moves 244
,, cited, 116, 132
Geographical conditions, misapprehensions regarding, 104, 106
,, conditions necessary for glaciation, 165
,, conditions part of Physical Theory, 93, 94
Geological facts in relation to modification of theory, 126
Georgia, South, why snow is perpetual, 51
Germany, mammoth remains found in, 186
,, glaciation of by Scandinavian ice-sheet, 169
Giescke, M., on Greenland, 69
Glacial epoch, condition of Europe at the time, 129, 132
,, three main causes of, 102
,, might be produced without very high eccentricity, 52
,, ice of, 246
,, how affected by geographical conditions, 165
,, dates of, derived from eccentricity, 174
Glacial periods of Tertiary age, 164
Glaciation not the result of eccentricity alone, 165
Glaciation, M, Woekiof on cause of, 57
Glacier-motion, heat transformed into, 256
,, regelation as a cause of, 248
Glen Brora, moraine of, 116
Glen Torsa, moraines on 30 feet beach, 116
Graham Land under glaciation, 113
Gravitation, theory of sun's heat irreconcilable with geological facts, 291

Gravitation alone cannot account for sun's heat, 299
,, alone cannot originate suns or nebulæ, 300
Gravity, a cause of descent of glaciers, 254
Greenland, heat received from the sun, 48
,, coldness of air during summer, 84
,, how its ice may be removed, 147
,, misapprehensions re-regarding, 81
,, why the summers are cold, 54
,, free from ice during any of the inter-glacial periods, 195
,, Nordenskjöld on absence of boulders in strata of, 170
,, rainfall and snowfall in, 237
,, ice of, 64
,, whale in Carse-deposits, 116
,, Rink, Heyes, Brown, &c., on, 64-66
,, glaciers, high velocity of, 239
Gulf Stream, influence in distribution of heat, 198
,, absolute amount of heat conveyed by, 146

HABENICHT, H., on German inter-glacial beds, 133
Haeckel, Prof., on age of the earth, 289, cited, 291
Hamilton, Sir William, definition of Cause, 103
Hampshire deposits, 163
Haughton, Prof., on change of axis of rotation, 4
,, cited, 294, 295
Heat of Aërial currents, 23
,, cut of by the atmosphere, 37
,, trapped," 45
,, rays "sifting" of, 44
Heat evolved by freezing, 45
,, as a cause of glacier motion, 257
,, absolute amount conveyed by Gulf-stream, 146
Hedenstrom, M., on the mammoth, 179, 181
Helix schrencki, 183
Helland, Mr. A., on German inter-glacial beds, 133
,, on Scandinavian ice-sheet, 133
,, on density of ice, 209
,, on rate of motion of Greenland ice, 239
,, on thickness of ice-sheet of Norway, 247
Helmholtz, theory of origin of sun's heat, 265
,, formula of condensation of sun's mass, 265
,, cited, 282, 290
Herschel, Sir W., cited, 309
Hessle boulder-clay of Mr. Wood, 131
,, not post-glacial, 132
Heyes, Dr., on Greenland ice, 64
High Land in relation to glaciation, 85
Hill, Rev. E., evaporation in relation to glaciation, 23
,, on change of axis of rotation, 5
,, answer to question, 57
,, objection considered, 46
,, cited, 48
Hippopotamus in England, 134
Hooker, Sir Joseph D., theory of Antarctic ice, 75
,, on tropical plants, 10
,, on Antarctic ice, 111
,, cited, 76
Hopkins, Mr., on internal heat, 2
Howorth, Mr., on the mammoth in Siberia, 178

Howorth, Mr., on mammoth in Europe, 186
,, on evidence from shells, 183
,, on plants of Canstadt, 189
,, cited, 180, 182, 187
Huggins, Mr., on nebula, 308
Hull, Prof., on great Irwell fault, 268
Hutton, Dr., on expansive force of ice in freezing, 215
Hypothesis, reason for avoiding, 26

ICE of Greenland and Antarctic continent, 64
,, at S. Pole, why it is thick, 78, 81
,, another reason why it does not melt, 123
,, permanent without perpetual snow, 89
,, melting point lowered by pressure, 215, 227
,, melting, resulting from mutual reaction of the physical causes, 122
,, paleochrystic, 77
,, cliffs of America, 192
,, how moved up hill slopes, 255
Icebergs, Antarctic, peculiar ice formation of, 70
,, testimony of, 207
,, found in Southern Ocean, 207
,, of Southern Ocean, enormous size of, 208
,, mean density of, 209
,, estimates of their thickness, 209
,, height of, determined by their mean density, 210
Ice-sheet, Antarctic, conditions of, 202
,, of north-western Europe, extent of, 246
,, interior of, warmer than surface, 217
,, explanation of passage across valleys, 255
Ice strata, variations is not due to pressure or melting, 231
,, thinning out of, due to dispersion, 232
Incandescence of Nebulous masses, how produced, 284
Indiga, port of, 182
Interglacial beds of Scotland, 130
,, of England, 131
,, of Germany, 133
,, of Switzerland, 133
Interglacial climate, characteristics of, 185
,, periods in Siberia, 184
Interglacial periods in Greenland, 195
,, of Pleistocene times, 127
,, interval between, 137
,, difficulty in detecting climatic character, 135
,, objection as to the number of, 137
,, probably five, 137
,, number detected in Germany, Denmark, &c., 138
,, shorter than Glacial, 141
,, less marked in temperate regions than Glacial, 141
Intertropical water, effect on polar climate, 165
Internal heat of the earth, 2

JAMIESON, Mr. on ice markings on Schiehallion, 212
Jenson, on Greenland ice, 64
Jentzsch, Dr., on German Interglacial beds, 133

KAMSKATKA, mammoth found there, 179
Kötzebue Sound, ice-cliffs of, 192
Krestowkoje, trees at, 183

LA CELLE, tufa of, 189
Land-connection with the Faröes, objection to, 199
,, between Europe, Faröes, and Greenland, 197
Land-ice theory, objections to considered, 246
Land-surface, absence of in Tertiary strata, 168
Langley, Prof., on temperature of space, 20
,, observations at Mount-Whitney, 20
,, remarkable conclusion by, 32
Le Coq., Prof. H., on change of climate, 3
Lena, river, mammoth found there, 179
Leverrier's formulæ of eccentricity, 38
Liachof Archipelago, mammoth found there, 179
Life on earth, testimony of geology as to age of, 266
,, age of, 299
Limax agrestis, 184
Limnæa auricularia, 183
Lockyer, Mr., on evolution of the planets, 307
,, cited, 309
Lyell, Sir Charles, on change of climate, 3
,, on geographical conditions, 97
,, on Mississippi, 275

M'CLURE, Capt., Arctic trees, 193
M'Farland, Prof., on eccentricity of earth's orbit, 38, 39
Mackenzie River, 192
Maak, M., shells found by, 183
Main characteristics of Interglacial climate, 185
Mammoth in Siberia, 178
,, ,, Europe, 186
,, Interglacial, 184
,, Glacial as well as Interglacial, 190
,, Dr. Rae, on, 191
Mars, why not under glaciation, 92
Mecham, Lieut., on Arctic trees, 193
Meech, cited, 48
Melville Island, wood found at, 193
Merklin, Prof., on limit of wood, 183
Meyer, Dr., meteoric theory, 265
Mild polar climates, 143
,, ,, Mr. A. R. Wallace on, 153-157
,, ,, due to intertropical water, 166
Miocene glacial deposits of Italy 172
,, period, date of, 174
Misconception, fundamental, 46
Mississippi, amount of sediment carried by, 275
Moisture in relation to glaciation, 85
Mollusca of the tufa, evidence from, 190
Montpellier, tufa of, 188
Moret, in the valley of the Seine, deposit near, 188, 190
Mosley, Canon, on insufficiency of gravity to shear ice, 248
Motion in space, 291
,, ,, primal cause of sun's light and heat, 287
Motion of stars, difficulty of estimating, 313
,, ,, our knowledge of, dependent on their directions, 313
Mount-Whitney expedition, 20
Mountains, carved out by denudation, 278
Mousson on the lowering of the melting point of ice, 215, 227

Mutual reaction of physical agents, 52, 54, 120
Mutual reaction of physical agents, misapprehension regarding, 52
Mytilus edulis at Spitzbergen, 194

NEBULÆ, probable origin of, 297
,, reason why they occupy so much space, 301
,, why of such various shapes, 304
,, why they emit such feeble light, 304
,, heat and light cannot result from condensation, 306
,, gaseous, first condition of, 308
,, irresolvability due to gaseous condition, 308
,, Prof. Tait's suggestion as to nebula, 310
Newcomb's, Prof., reason for considering his objections, 18
,, objections to my terms, 19
,, on aërial currents, 23, 24
,, on Prevost's theory of exchanges, 31
Neumayer, Dr., on drift currents, 111
Newton's law of radiation, 22
,, in relation to temperature, 145
"Noashina," 181, 193
Nordenskjöld, Prof., on absence of boulders in Greenland strata, 170
,, on Spitzbergen interglacial bed, 194
,, on interglacial deposits, 194
Northward-flowing currents, causes affecting, 112
Northern Siberia, mammoth in, 178
,, former limit of wood in, 180
North-Western Europe, path of ice in, 133
Norway, ice-sheet of, 247

OBJECTIONS to theory of stellar heat and light considered, 310
Obliquity of ecliptic, change of, 4, 115
Ocean-currents, influence of, in Arctic regions, 8
Orkney Islands, glaciation of, 169
Osborn, Capt., on Arctic trees, 193

PALÆONTOLOGY in relation to modification of theory, 126
Paleochrystic sea, 77
Parallax of stars, few known, 314
Payer, Commander Julius, on Franz Josef Land, 76
,, on permanent ice at sea level, 76
Peach, Mr. B. N., on great fault between Silurian and O.R. sandstone, 269
Penck, Dr. A., on German interglacial beds, 133
Perihelion summer not hot, 53
Permanent ice without perpetual snow, 89
Physics in relation to Mr. Wallace's modification, 100
Physical conditions of Antarctic ice, 202
Physical agents in relation to melting of ice, 119
,, mutual reaction of, 52, 119
Physical theory, no hypothesis, 16
,, three factors of, 84
,, modification examined, 89
,, general statement of, 91
,, does not account for all conditions, 91
,, a misapprehension regarding, 92

Physical theory, why so named, 92
,, recognises geographical conditions, 93
,, points of agreement, 95-97
Pisidium fontinale, 184
Planets, elements of, 308
Planorbis albis, 183
Polar regions, mild climates exceptional in, 171
Poisson, M., on cause of geological climate, 2
Prevost's theory of exchanges, 31
Prince Patrick's Island, wood found at, 193
Provence, tufa of, 188
Pouillet, M., on temperature of space, 259
Pouillet and Herschel on temperature of space, 144
Poles, heat received to that at the Equator, 143
Polar regions, glaciation, normal condition, 171

RADIATION of a particle, 22
,, of heat through plates, Professor Balfour Stewart on, 261
,, effect of, on surface of ice-sheets, 217
Rae, Dr., on mammoth in Siberia, 191
Raised beach with moraine matter on, 116
Ramsay, Prof, A. C., on great faults in North Wales, 268
Reaction of the physical agents, 54, 120
Reade, Mr. Mellard, cited, 279
Regelation, three theories of, 249
,, theory of, from 'Climate and Time,' 251
"Rejoinder," Prof. Newcomb's, 31, 32, 34
Rhinoceros tichorhinus, 180
Rink, Dr., on Greenland ice, 64
,, on rainfall and snowfall in Greenland, 237
Rink, Dr., cited, 48
Rock-basins, how excavated by land ice, 254
Rogers, Prof. H. D., on fault in the Appalachians, 270
Ross, Sir John, on snowfall of Antarctic regions, 79
,, on absence of water on Artarctic ice-sheet, 218
Ruprecht, Herr von, on Siberia, 182
Russia, mammoth remains found in, 186

SANDWICH Land under glaciation, 113
Sannikow on former climate of Siberia, 182
Saporta, M., on tufa, 188, 189
Schmidt on deposits on the Tundra, 182
Scoresby, Capt., on Greenland, 69
,, on coldness of Greenland air, 84
Scotland, inter-glacial beds of, 130
Sea, temperature higher than that of land, 26
,, mean temparature, why it is high? 26-31
Sea-level, oscillations of, 139, 140
Secular cooling of the earth, Sir W. Thomson on, 295
Shaler, Prof., theory of Antarctic ice, 76
,, on Antarctic ice, 111
Shells in Tufa, 190
Siberia, why snow is not perpetual, 88
,, mammoth in, 178
,, inter-glacial period of, 190
,, North warmer during mammoth epoch, 180
,, wood found in, 181
Siemens, Dr., theory of the sun, 298
"Sifting" of rays, 44
Skeletons of mammoth in Europe, 186

Smithers, Capt., measurement of icebergs by, 210
Snow, misapprehensions regarding, 40
,, influence of, 40
,, amount melted, 41
,, influence on climate, 41, 45
,, melting not proportional to heat received, 46-55
,, melting misconception regarding, 47
,, conservation of, 46
,, permanent in northern hemisphere at sea level, 76
,, a permanent source of cold, 84
,, perpetual without high lands, 87
Snow-fall, why great at south pole, 79
,, in relation to glaciation, 85
South Georgia, heat received from sun, 44, 48
,, under glacial conditions, 113
South Pole, why the ice must be thick there, 78—81
,, probable thickness of ice at, 227, 241, 245
,, ice at below freezing point, 227
,, greatest thickness of ice there independent of amount of snowfall, 236
,, mode of dispersion of ice from, 233
South-polar ice-cap, Mr. Wallace on, 78
South Shetland under glaciation, 113
Southern Europe, deposits of, 188
Southern hemisphere warmer than northern, 33
Space, temperature of, an important factor, 19
,, Prof. Langley on temperature of, 20
,, Pouillet and Herschel on temperature of, 20, 21
,, temperature of, 258
Spectrum analysis of stellar motions, on, 313
Spencer, Mr. Herbert cited, 288
Spitzbergen, interglacial beds of, 194
Star-clusters, formation of, 309
Stars of great velocities, why not observed, 312
Stellar space, on bodies moving in, 286
Stellar bodies, motions of, converted into heat by collision, 301
Stewart, Prof. Balfour, on radiation, 22
,, on radiation through plates, 261
Stockwell, Mr., on obliquity of ecliptic, 4
,, formulæ of eccentricity, 39
Stoney, Mr. Johnstone, cited, 287, 288
Stratified rocks, age of, as determined by denudation, 277
,, age of, 278
,, evidence as to age of earth, 267
Submergences, oscillations of sea-level in relation to, 139
,, objection as to the number of, 139
,, number unknown, 139
Succinea putris, 184
Sun, minimum age of, determined by that of the earth, 280
,, loss of heat, estimate of, 290
,, probable origin and age of, 282
,, how original temperature obtained, 283
,, nebulous mass of, how superheated, 283
Sun's heat, origin and age of, 264
,, as derived from gravitation, 265
Surface temperature of Antarctic ice, 225
Sweden, mammoth remains found at, 186
Switzerland, interglacial beds, 133

Switzerland, "*Flysch*" of, 173
,, illustration from ice of, 245
,, glaciers, temperature determined by pressure, 222

TAIT, Prof., on age of the sun, 266.
,, on nebula, 310
Tay Valley, Carse-clays of, 115
Temperature of space, bearing on terrestrial physics, 258
,, Prof. Langley on, 20
,, an important factor, 19
,, mean of ocean, reason why it is high, 26-31
,, of ocean greater than that of land, 26
Temporary stars, why so rare, 315
,, causes of, 315
Terrestrial physics, most important problems in, 1
Tertiary period, influence of eccentricity during, 157
,, as affected by eccentricity, 160
,, evidence of glaciation in, 172
Tertiary fossil floras, Mr. J. Starkie Gardner on, 161
,, glacial epochs, 164
Thermodynamics, misapprehensions regarding, 33
Thomson, Prof. James, on cause of regelation, 250
,, on lowering of melting point of ice, 215, 227
,, cited, 203
,, formula, 227
Thomson, Sir William, on internal heat, 2
,, theory of several climates, 6
,, on mild Arctic climates, 148
Thomson, Sir William, on age of earth, 292
,, cited, 14
,, on underground heat, 213
Thomson, Sir Wyville, on Antarctic regions, 69-73, 203
,, on Antarctic icebergs, 70-74
,, on thickness of Antarctic ice-sheet, 225
,, on thickness of icebergs, 209
,, on stratification of Antarctic ice, 229
,, on Antarctic ice-cap, 232
Tidal retardation, argument from, 292
Torrell, Prof., on German interglacial beds, 133
Tournonër, M., on shells, 190
Towson, Mr., on icebergs of Southern Ocean, 207
,, cited, 211
Tropical and Arctic floras, Mr. J. S. Wood on commingling of, 162
Tufa, Prof. J. Geikie on, 187
,, Mr. Howorth on, 188, 189
,, of Provence, 188; of Moret, 188; of Canstadt, 189; of Le Celle, 189
,, of Europe, 187
,, mollusca of, 190
,, of Tuscany, 188
Twisden, Rev. J. F., on changes of axis of rotation, 5
,, cited, 294
Tyndall, Prof., experiments on passage of heat through ice, 252
,, experiments on heat-absorbing power of aqueous vapour, 260

UNDERGROUND heat, 214
,, ,, in relation to Antarctic ice-sheet, 225
Unio littoralis, 131, 134

VALVATA *cristata*, 183
,, *pisinatis*, 183
Velocity of bodies in space, 205

WAHNSCHAFFE, F., on German interglacial beds, 133
Wallace, A. R., on heat evolved by freezing, 45
,, on south polar ice-cap, 78
,, on effect of winter in aphelion, 83
,, on storage of cold, 83
,, points of agreement with, 95-97
,, on influence of a land-barrier, 98
,, modification of theory, 100
,, on present effect of eccentricity, 112
,, on reaction of physical agents, 121
,, on why the ice does not melt, 123
,, on interglacial beds of Scotland, 130
,, on interglacial beds of England, 131
,, on mild Arctic climates, 153
,, on table of eccentricity, 167
,, cited, 279
Whale in Garse-deposits, 116
Wilkes, Lieut., estimate of snowfall in Antarctic regions, 237
,, on snowfall of the Antarctic regions, 79
Winter, why explanation begins with, 57
Woeikof, M., on the cause of glaciation, 57
Wood, evidence from, 180
Wood, Searles V., on fogs, 45
,, on interglacial beds of England, 131
,, theory of the commingling of tropical and Arctic floras, 162
,, on absence of Glacial Epochs in tertiary strata, 167
Wrangell, on remains of the mammoth, 179
,, on trees in Siberia, 181, 182

YENISSEI, mammoth found there, 179, 181
Yukagirs, country of, mammoth found there, 179

ZERO, absolute of temperature, 21
Zones, quantity of heat possessed by each, 151

THE END.

PRINTED BY WM. HODGE & CO., 123 HOPE STREET, GLASGOW.

Printed in the United States
By Bookmasters